Additional
Science
(AQA)

Complete Revision and Practice

Nigel Saunders

Published by BBC Active, an imprint of Educational Publishers LLP, part of the Pearson Education Group Edinburgh Gate, Harlow, Essex CN20 2JE, England

Text copyright © Nigel Saunders 2008
Design & concept copyright © BBC Active 2008, 2010

BBC logo © BBC 1996. BBC and BBC Active are trademarks of the British Broadcasting Corporation

First published 2008
This edition 2010

ISBN 978-1-4066-5435-6

Printed in China CTPSC/01

The Publisher's policy is to use paper manufactured from sustainable forests.

D0582685

Minimum recommended system requirements
PC: Windows(r), XP sp2, Pentium 4 1 GHz processor (2 GHz for Vista), 512 MB of RAM (1 GB for Windows Vista), 1 GB of free hard disk space, CD-ROM drive 16x, 16 bit colour monitor set at 1024 x 768 pixels resolution
MAC: Mac OS X 10.3.9 or higher, G4 processor at 1 GHz or faster, 512 MB RAM, 1 GB free space (or 10% of drive capacity, whichever is higher), Microsoft Internet Explorer® 6.1 SP2 or Macintosh Safari™ 1.3, Adobe Flash® Player 9 or higher, Adobe Reader® 7 or higher, Headphones recommended

If you experiencing difficulty in launching the enclosed CD-ROM, or in accessing content, please review the following notes:
1 Ensure your computer meets the minimum requirements. Faster machines will improve performance.
2 If the CD does not automatically open, Windows users should open 'My Computer', double-click on the CD icon, then the file named 'launcher.exe'. Macintosh users should double-click on the CD icon, then 'launcher.osx'
Please note: the eDesktop Revision Planner is provided as-is and cannot be supported.
For other technical support, visit the following address for articles which may help resolve your issues:
http://centraal.uk.knowledgebox.com/kbase/

If you cannot find information which helps you to resolve your particular issue, please email: Digital.Support@pearson.com.
Please include the following information in your mail:
- Your name and daytime telephone number.
- ISBN of the product (found on the packaging.)
- Details of the problem you are experiencing - e.g. how to reproduce the problem, any error messages etc.
- Details of your computer (operating system, RAM, processor type and speed if known.)

Contents

Exam board specification map   iv

Introduction   vi

Topic checker   x

Topic checker answers   xvii

Biology

Cells	2
Diffusion and osmosis	4
Photosynthesis	6
Minerals	8
Food chains and pyramids	10
Food production	12
The carbon cycle	14
Respiration and enzymes	16
Digestion	18
Uses of microorganisms and enzymes	20
Maintaining internal conditions	22
Diabetes	24
DNA	26
Cell division	28
Inheritance	30
Inherited disorders	32
Inheritance (Higher Tier)	34

Chemistry

Atomic structure	36
Ions and ionic bonds	38
Simple molecules	40
Structure and properties of materials	42
Structure and bonding (Higher Tier)	44
Relative atomic mass	46
Formulae and percentage composition	48
Atom economy and balancing equations	50

Chemical calculations (Higher Tier)   52

Rates of reaction   54

Reversible reactions   56

The Haber process   58

Electrolysis   60

Acids, bases and neutralisation   62

Making salts   64

Physics

Speed, distance and acceleration   66

Forces   68

Weight and falling   70

Stopping distances   72

Work and kinetic energy   74

Momentum   76

Static electricity   78

Electrical circuits   80

Resistance   82

Parallel and series circuits   84

Mains electricity   86

Power   88

Energy and charge (Higher Tier)   90

Radiation   92

Fusion and fission   94

Exam questions   96

Chemistry data   107

Answers to exam questions   109

Complete the facts   *

Complete the facts answers   *

Answers to practice questions   112

Glossary   116

Web links   *

Last-minute learner   119

* Only available in the CD-ROM version of the book.

Exam board specification map

This is a single award GCSE, separate from and taken after or at the same time as Science A or B. This award together with an award in GCSE Science provides the nearest equivalent to the previous GCSE Science: Double Award. The content follows on from that of GCSE Science. However, the emphasis of this specification, and the three separate sciences, Biology, Chemistry and Physics, is somewhat different. Whereas Science A and B emphasise evaluating evidence and the implications of science for society, this specification has a greater emphasis on explaining, theorising and modelling in science.

There are three 45 minute written papers with structured questions, one paper for each of Biology 2, Chemistry 2 and Physics 2, available in January and June.

Topics	AQA Additional Science
Biology 2	
Cells	✓
Diffusion and osmosis	✓
Photosynthesis	✓
Minerals	✓
Food chains and pyramids	✓
Food production	✓
The carbon cycle	✓
Respiration and enzymes	✓
Digestion	✓
Use of microorganisms and enzymes	✓
Maintaining internal conditions	✓
Diabetes	✓
DNA	✓
Cell division	✓
Inheritance	✓
Inherited disorders	✓
Inheritance (Higher Tier)	✓
Chemistry 2	
Atomic structure	✓
Ions and ionic bonds	✓
Simple molecules	✓
Structure and properties of materials	✓
Structure and bonding (Higher Tier)	✓
Relative atomic mass	✓
Formulae and percentage composition	✓
Atom economy and balancing equations	✓
Chemical calculations (Higher Tier)	✓

Topics	AQA Additional Science
Rates of reaction	✓
Reversible reactions	✓
The Haber process	✓
Electrolysis	✓
Acids, bases and neutralisation	✓
Making salts	✓
Physics 2	
Speed, distance and acceleration	✓
Forces	✓
Weight and falling	✓
Stopping distances	✓
Work and kinetic energy	✓
Momentum	✓
Static electricity	✓
Electrical circuits	✓
Resistance	✓
Parallel and series circuits	✓
Mains electricity	✓
Power	✓
Energy and charge (Higher Tier)	✓
Radiation	✓
Fusion and fission	✓

Introduction

How to use GCSE Bitesize Complete Revision and Practice

Begin with the CD-ROM. There are five easy steps to using the CD-ROM – and to creating your own personal revision programme. Follow these steps and you'll be fully prepared for the exam without wasting time on areas you already know.

Topic checker

Step 1: Check

The Topic checker will help you figure out what you know – and what you need to revise.

Revision planner

Step 2: Plan

When you know which topics you need to revise, enter them into the handy Revision planner. You'll get a daily reminder to make sure you're on track.

Revise

Step 3: Revise

From the Topic checker, you can go straight to the topic pages that contain all the facts you need to know.

Web Bite

- Give yourself the edge with the WebBite buttons. These link directly to the relevant section on the BBC Bitesize Revision website.

Audio Bite

- AudioBite buttons let you listen to more about the topic to boost your knowledge even further. *

Step 4: Practise

Check your understanding by answering the Practice questions. Click on each question to see the correct answer.

Exam Bite

Step 5: Exam

Are you ready for the exam? ExamBite buttons take you to an exam question on the topics you've just revised. *

* Not all subjects contain these features, depending on their exam requirements.

You can choose to go through every topic from beginning to end by clicking on the Interactive book and selecting topics on the Contents page.

 Find all of the exam questions in one place by clicking on the Exam questions tab.

 The Last-minute learner gives you the most important facts in a few pages for that final revision session.

 You can access the information on these pages at any time from the link on the Topic checker or by clicking on the Help button. You can also do the Tutorial which provides step-by-step instructions on how to use the CD-ROM and gives you an overview of all the features available. You can find the Tutorial on the Home page when you click on the Home button.

Other features include:

 Click on the draw tool to annotate pages. N.B. Annotations cannot be saved.

 Click on Page turn to stop the pages turning over like a book.

 Click on the Single page icon to see a single page.

 Click on this arrow to go back to the previous screen.

Click on Contents while in the Interactive book to see a contents list in a pop-up window.

Click on these arrows to go backward or forward one page at a time.

 Click on this bar to switch the buttons to the opposite side of the screen.

Click on any section of the text on a topic page to zoom in for a closer look.

N.B. You may come across some exercises that you can't do on-screen, such as circling or underlining, in these cases you should use the printed book.

About this book

Use this book whenever you prefer to work away from your computer.
It consists of two main parts:

 A set of double-page spreads, covering the essential topics for revision from each area of the curriculum. Each topic is organised in the following way:

- a summary of the main points and an introduction to the topic

- lettered section boxes cover the important areas within each topic

- key facts highlighting essential information in a section or providing tips on answering exam questions

- practice questions at the end of each topic to check your understanding.

A number of special sections to help you consolidate your revision and get a feel for how exam questions are structured and marked. These extra sections will help you to check your progress and be confident that you know your stuff. They include:

- Topic checker – quick questions covering all topic areas

- exam-style questions and answers similar to the ones you will meet in AQA's GCSE Additional Science module examinations. Suitable for Foundation Tier and Higher Tier candidates

- Complete the facts – check that you have the most important ideas at your fingertips

- Last-minute learner – the most important facts in just a few pages.

About your exam

Get organised

You need to know when your exams are before you make your revision plan. Check the dates, times and locations of your exams with your teacher, tutor or school office.

On the day

Aim to arrive in plenty of time, with everything you need: several pens, pencils and a ruler.

On your way, or while you're waiting, read through your Last-minute learner.

In the exam room

When you are issued with your exam paper, you must not open it immediately. However, there are some details on the front cover that you can fill in before you start the exam itself (your name, centre number, etc.). If you're not sure where to write these details, ask one of the invigilators (teachers supervising the exam).

When it's time to begin writing, read each question carefully. Remember to keep an eye on the time.

Finally, don't panic! If you have followed your teacher's advice and the suggestions in this book, you will be well prepared for any question in your exam.

Picking up marks in exams

The examiners are not trying to trick you. They just want to give you the chance to show what you can do. But they are not mind readers. They can only mark what you actually do on the examination paper. Here are some general tips:

- Read through the whole paper for a few minutes at the start and tick the easiest questions as you go.

- Answer your ticked questions first. This will help you feel more calm and confident.

- Tackle the remaining questions.

- Remember to read the questions carefully and underline important words. The information at the start is there to help you, so don't ignore it.

More advice for written exams

Keeping to time
- The written exams work out at one mark per minute on average. You are likely to run out of time if you spend ten minutes on a question worth just one mark.

- The amount of space and the number of marks available are clues to how much to write. You probably won't have written enough if you write one line when there are four marks and six lines.

Calculations
- Write down the equation you need to answer the question.

- Show your working out.

- Show the units if asked.

- Check that the answer makes sense.
  For example, is a snail really likely to move at 100 m/s on its own?

Graphs
- Use a sharp pencil.

- You are likely be given a partly completed graph to finish in the Foundation Tier exam. In the Higher Tier exam you are likely to have to draw your own axes. Try to use more than half the area of the graph paper without choosing a strange scale.

- Label both axes and include the units.

- Plot each point with a neat cross, or use a ruler to draw bars.

- In a bar chart, don't waste time shading in the bars unless the question asks you to. In a line graph, draw a line or smooth curve with a single stroke of your pencil.

Topic checker

Go through these questions after you've revised a group of topics, putting a tick if you know the answer.

You can check your answers on pages xvii–xxi.

Biology

>> Cells

1	Which part of the cell controls the movement of substances into and out of the cell?	☐
2	What happens in the ribosomes?	☐
3	State three components that are found in plant cells but not in animal cells.	☐
4	What is diffusion?	☐
5	Why does a piece of potato shrink when placed in very salty water?	☐

>> Plants

6	What are the reactants in photosynthesis?	☐
7	What is the function of chlorophyll?	☐
8	Name the storage substance in plants, formed from glucose.	☐
9	State one reason why plants need nitrate ions, and outline the effects of nitrate ion deficiency in plants.	☐
10	State one reason why plants need magnesium ions, and outline the effects of magnesium ion deficiency in plants.	☐

>> Chains and cycles

11	What information does a pyramid of biomass provide?	☐
12	State two reasons why less energy is contained in the biomass of each successive stage in a food chain.	☐
13	Which process removes carbon dioxide from the atmosphere in the carbon cycle?	☐
14	State two processes that return carbon dioxide to the atmosphere in the carbon cycle.	☐

>> Respiration

15	What is the word equation for aerobic respiration?	
16	Where does most respiration take place in the cell?	
17	State three uses for the energy released by respiration.	

>> Enzymes

18	What is an enzyme?	
19	Explain why enzymes stop working at high temperatures.	
20	Why do biological detergents contain lipase?	
21	Outline what isomerase does, and why it is useful in the food industry.	

>> Digestion

22	What is digestion?	
23	State three sources of protease in the gut.	
24	Where is bile produced, and what does it do?	

>> Maintaining internal conditions

25	State three conditions in the body, apart from temperature, that must be controlled.	
26	What is the function of the thermoregulatory centre?	
27	What is urea, and where is it produced?	
28	Outline two ways in which the body temperature can be reduced if it gets too high.	

Topic checker

>> Cell division and inheritance

29 What type of cell division gives rise to identical cells?

30 What is a gene?

31 What sex chromosomes are found in human females, and in human males?

32 Outline why cystic fibrosis can be passed on by parents who do not have the disorder.

Chemistry

>> Atoms and electrons

33 What is the atomic number of an atom?

34 How are the elements arranged in the modern periodic table?

35 What is the link between the number of electrons in the highest energy levels and groups in the periodic table?

36 An element from group 7 is placed in the third period of the periodic table. What is its electronic structure?

>> Bonding

37 In general, what charge do metal ions have, and what charges do non-metal ions have?

38 The electronic structure of a magnesium atom is 2,8,2. What is the electronic structure of a magnesium ion Mg^{2+}?

39 What is an ionic bond?

40 What is a covalent bond?

41 Explain why water is made of simple molecules but diamond is made of giant molecules.

42 Outline the differences between the structures of diamond and graphite.

>> Properties of different substances

43   Why do metals and graphite conduct electricity well?

44   Outline how you could distinguish between a substance made of ions and one made of simple molecules.

45   Why do ionic compounds conduct electricity when molten or dissolved in water, but not when they are solid?

46   What are nanoparticles? How do their properties compare to the properties of the same substance in bulk?

>> Relative masses

47   An atom has the symbol $^{23}_{11}$Na. What information do the numbers give you about the atom?

48   If the mass of a proton is 1, what is the mass of a neutron, and the mass of an electron?

49   What are isotopes of an element?

50 $A_r(H) = 1$; $A_r(O) = 16$; $A_r(Mg) = 24$. Work out the M_r of magnesium hydroxide, $Mg(OH)_2$.

51   The M_r of sodium hydroxide NaOH is 40. If the A_r of oxygen O is 16, what is the percentage by mass of oxygen in NaOH?

>> Rate of reaction

52   In general, how can the rate of a reaction be measured?

53   State four ways in which the rate of a chemical reaction can be increased.

54   What is the activation energy?

55   Outline why increasing the concentration of a reactant in solution increases the rate of reaction.

Topic checker

>> Reversible reactions

56 What is a reversible reaction? ☐

57 The forward reaction in a certain reversible reaction is exothermic. What can you say about the reverse reaction? ☐

58 State the typical conditions used in the Haber process. ☐

>> Reactions in solution

59 State three products of the electrolysis of sodium chloride solution. ☐

60 In the electrolysis of molten lead bromide, which electrode are lead ions attracted to? Are the lead ions reduced or oxidised there? ☐

61 What is the symbol equation for neutralisation? ☐

62 Which salt is made when potassium hydroxide solution reacts with sulfuric acid? ☐

>> Higher Tier only

63 The expected yield of a certain reaction is 50 g. If the actual yield is 45 g, what is the percentage yield? ☐

64 Ethene reacts with steam to make ethanol: $C_2H_4 + H_2O \rightarrow C_2H_5OH$. What is the atom economy of this process? ☐

65 What effect will increasing the pressure have on the position of equilibrium in this reversible reaction? $2SO_2(g) + O_2(g) \rightarrow 2SO_3(g)$. ☐

66 Balance this half-equation: $H^+ + e^- \rightarrow H_2$ ☐

Physics

>> Forces and motion

67 An object travels 25 m in 10 s. What is its speed? ☐

68 What information does the velocity of an object provide? ☐

69 What information does the area under a velocity-time graph provide? ☐

70 State the equation used to work out the acceleration of an object from information about velocity and time. ☐

71 What information does the slope of a velocity-time graph provide? ☐

72   Explain why an object falling through the air may reach a terminal velocity. ☐

73   An object with a mass of 2 kg is accelerated by 2 m/s^2. What resultant force was applied to the object? ☐

74   State four factors that could increase the stopping distance of a car. ☐

75   A resultant force of 5 N is applied to an object and it moves 20 m in the direction of the force. Calculate the work done. ☐

76   What can be said about the momentum in a collision or explosion if there are no external forces acting? ☐

>> Static electricity

77   An object becomes positively charged. Has it gained or lost electrons? ☐

78   Explain why a party balloon is attracted to your hair after you rub it against your head. ☐

79   What happens to a charged object when it is connected to earth? ☐

>> Electrical circuits

80   What are the differences between the symbol for a resistor, the symbol for a fuse, and the symbol for a variable resistor? ☐

81   A graph of current against potential difference has a straight line. Is the component involved most likely to be a resistor at constant temperature, a filament lamp or a diode? ☐

82   In what way does the resistance of an LDR depend on light intensity? ☐

83   Calculate the resistance of a component when 2 A flows through it with a potential difference of 10V. ☐

84   Four 1.5V cells are connected in series. One faces the opposite way to the other three. What is the potential difference provided? ☐

85   What can we say about the potential difference across components connected in parallel? ☐

86   State the voltage and frequency of mains electricity. ☐

87   How does a fuse work? ☐

88   A device transforms 10 000 J in 100 s. What is its power? ☐

89   Calculate the current flowing through a 2.3 kW electric fire at 230V. ☐

Topic checker

>> Radiation

90   State three natural sources of background radiation.

91   Outline the difference between nuclear fusion and nuclear fission.

92   Outline how a nuclear chain reaction works.

93   Name an isotope of uranium or plutonium commonly used as a nuclear fuel.

>> Higher Tier only

94   Calculate the kinetic energy of an object with a mass of 4 kg moving at 4 m/s.

95   What is the relationship between force, momentum and time?

96   When can you get a spark if a charged object is brought near to an earthed conductor?

97   What is the relationship between charge, current and time?

98   100 C of charge flows with a potential difference of 10V. How much energy is transformed?

Topic checker answers

>> Cells

1	cell membrane
2	protein synthesis
3	in any order: cell wall, chloroplast, vacuole
4	the spreading of particles from a region of high concentration to a region of low concentration
5	Water is lost from the cells by osmosis.

>> Plants

6	water and carbon dioxide
7	It absorbs light energy for photosynthesis.
8	starch
9	for producing amino acids. Deficiency leads to stunted growth.
10	for producing chlorophyll. Deficiency leads to yellow leaves.

>> Chains and cycles

11	The mass of living material at each stage in a food chain (the width of each bar is proportional to the biomass)
12	two from: energy lost in waste, respiration, movement, heat loss
13	photosynthesis
14	two from: respiration by animals, respiration by plants, respiration by microorganisms, combustion of fuels

>> Respiration

15	glucose + oxygen $\rightarrow$ carbon dioxide + water (+ energy)
16	in the mitochondria
17	three from: building up small molecules into larger ones, muscle contraction, keeping warm, making amino acids in plants

>> Enzymes

18	a biological catalyst; a protein that increases the rate of reactions
19	Their shape changes and they become denatured.

Topic checker answers

| 20 | Lipases break down fats and oils in stains into fatty acids and glycerol, which are soluble and more easily washed out of clothes. |

| 21 | Isomerase converts glucose syrup into fructose syrup; this is very sweet so it can be used in small amounts in slimming foods. |

>> Digestion

| 22 | the break down of large molecules into smaller molecules |

| 23 | stomach, pancreas, small intestine |

| 24 | produced in the liver. It neutralises the acid produced by the stomach. |

>> Maintaining internal conditions

| 25 | water content, ion content, blood glucose concentration |

| 26 | monitor and control body temperature |

| 27 | a substance produced in the liver by the break down of excess amino acids |

| 28 | sweating and dilation of blood vessels supplying the skin capillaries |

>> Cell division and inheritance

| 29 | mitosis |

| 30 | a small section of DNA (which codes for a particular protein) |

| 31 | females are XX; males are XY |

| 32 | Cystic fibrosis is caused by a recessive allele. Parents who carry one of these alleles do not have the disorder, but there is a 25% chance of producing a child with two copies of the allele. |

>> Atoms and electrons

| 33 | the number of protons in the nucleus |

| 34 | in order of increasing atomic number |

| 35 | The number of electrons in the highest energy level is the same as the group number. |

| 36 | 2,8,7 |

>> Bonding

| 37 | Metal ions have positive charges and non-metal ions have negative charges. |

| 38 | $(2,8)^{2+}$ |

39 (electrostatic) force of attraction between oppositely charged ions

40 a shared pair of electrons

41 Water molecules each contain three atoms covalently bonded together, but diamond contains very many atoms covalently bonded together.

42 Diamond: each carbon atom is bonded to four others and no delocalised electrons. Graphite: each carbon atom is bonded to three others forming sheets with weak forces between them, and delocalised electrons in the layers.

>> Properties of different substances

43 They have free electrons (delocalised electrons).

44 Ionic compounds have high melting and boiling points, and conduct electricity when molten or dissolved in water. Simple molecular substances have low melting and boiling points, and do not conduct electricity.

45 Their ions are free to move.

46 particles between 1 nm and 100 nm in size. Their properties are different.

>> Relative masses

47 Its mass number is 23 and its atomic number is 11 (it has 11 protons, 11 electrons, and 12 neutrons).

48 Mass of neutron is 1. Mass of electron is very small

49 atoms with the same number of protons but different numbers of neutrons

50 58

51 40%

>> Rate of reaction

52 Measure the rate of disappearance of a reactant, or the rate of appearance of a product.

53 four from: increase the temperature, concentration of dissolved reactant, pressure of reacting gas, surface area of reacting solid, add a catalyst

54 the minimum energy needed for particles to react

55 frequency of collisions increases

>> Reversible reactions

56 A reaction in which the products can break down to form the original reactants

57 It is endothermic.

Topic checker answers

| 58 | 450 °C, 200 atmospheres, iron catalyst |

>> Reactions in solution

59	hydrogen, chlorine, sodium hydroxide solution
60	the negative electrode. They are reduced.
61	$H^+(aq) + OH^-(aq) \rightarrow H_2O(l)$
62	potassium sulfate

>> Higher Tier only

63	90%
64	100% (no other product is made)
65	It will move to the right.
66	$2H^+ + 2e^- \rightarrow H_2$

>> Forces and motion

67	2.5 m/s
68	the change in distance with time in a particular direction
69	distance travelled
70	acceleration = change in velocity ÷ time
71	acceleration
72	Its weight becomes balanced by air resistance so the resultant force is zero.
73	$2 \times 2 = 4$ N
74	four from: condition of tyre, condition of brakes, condition of road (e.g. icy, wet, dry), weather, condition of driver (tiredness, whether they have had alcohol or drugs), speed of the car, mass of the car
75	$5 \times 20 = 100$ J
76	It is conserved (stays the same).

>> Static electricity

77	lost electrons
78	The balloon and hair have opposite charges, so they attract each other.
79	It is discharged.

>> Electrical circuits

80	All three show a rectangle, but the fuse has a horizontal line through it and the variable resistor has a diagonal arrow through it.
81	a resistor at constant temperature
82	It decreases as the light intensity increases.
83	$10 \div 2 = 5\,\Omega$
84	3.0V
85	It is the same.
86	230V, 50Hz
87	When the current flowing through it becomes too great, the fuse wire melts and breaks the circuit.
88	$10\,000 \div 100 = 100\,\mathrm{W}$
89	$2300 \div 230 = 10\,\mathrm{A}$

>> Radiation

90	any three natural sources of background radiation, e.g. cosmic rays, UV light from the Sun, rocks, food, drink, radon gas
91	Nuclear fusion: smaller nuclei fuse to make a large nucleus. Nuclear fission: larger nucleus splits to make smaller nuclei.
92	A neutron causes a nucleus to split, releasing two or more neutrons. Each of these splits a nucleus, releasing even more neutrons. An increasing number of nuclei split each time.
93	Uranium-235 or plutonium-239

>> Higher Tier only

94	$\frac{1}{2} \times 4 \times 4^2 = 32\,\mathrm{J}$
95	force = change in momentum ÷ time taken for change
96	When the potential difference is high
97	charge = current × time
98	$10 \times 100 = 1000\,\mathrm{J}$

Cells

- All cells have some common features but plant cells have some extra features.

- Many cells are specialised to carry out particular functions.

A Animal cells

>> **key fact** Human cells and other animal cells have these parts:

part	function
cell membrane	controls the movement of substances into and out of the cell
nucleus	contains the cell's genetic material and controls the activities of the cell
cytoplasm	the site of most of the cell's chemical reactions, which are controlled by enzymes
mitochondria	release energy by respiration
ribosomes	where proteins are made (protein synthesis)

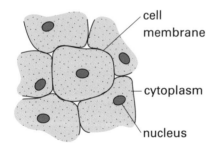

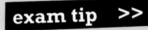

exam tip >>

Take care not to describe the nucleus as the brain of the cell.

Animal cells. Plant cells also have these features.

B Plant cells

① key fact Plant cells have the same parts as animal cells, but they also have a cell wall. This gives the cell its shape and strengthens the cell.

② key fact Plant cells also may have:
- Chloroplasts. These contain chlorophyll, a green protein that absorbs light energy to make food by photosynthesis.

- A permanent vacuole. This is filled with a watery cell sap. When the vacuole is filled, it pushes the cytoplasm against the cell wall. This helps the cell keep its shape.

cell membrane
cell wall
chloroplast
cytoplasm
vacuole
nucleus

Plant cells. Note their regular shape compared to the animal cells.

C Structure and function

key fact Cells may look different because different types of cell have different functions. Their structure helps them carry out their functions efficiently.

exam tip >>

You may be given some information, such as a diagram, to help you explain how the structure of a cell is related to its function.

Here are some examples:

type of cell	structure	function	
nerve cell	long and thin	carries nerve impulses from one part of the body to another	
sperm	long tail	allows the sperm to move towards an egg cell	
root hair cell	large surface area	absorbs water and dissolved minerals	
leaf cell	box shape with many chloroplasts	absorbs light energy for photosynthesis	

>> practice questions

1 List:

   a) the feature common to all cells,

   b) the features found only in plant cells.

2 Suggest why a plant wilts (becomes floppy) when it is short of water.

3 What substances control chemical reactions in the cytoplasm?

Diffusion and osmosis

 Diffusion and osmosis are processes that allow dissolved substances to move into and out of cells.

A Diffusion

1 **key fact** Diffusion is the movement of a substance from a region where it is more concentrated to a region where it is less concentrated.

2 **key fact** The diffusing substance moves down a concentration gradient. The greater the difference in concentration, the faster the rate of diffusion.

3 **key fact** Diffusion works for gases and substances in solution.

4 Dissolved substances can move into and out of cells because of diffusion.

5 Gases are exchanged at the surface of the lungs because of diffusion:

- oxygen moves from the air into the blood because it is at a higher concentration in the air

- carbon dioxide moves from the blood into the air in the lungs because it is at a higher concentration in the blood.

network of small blood vessels surrounding the alveoli and exchanging gases with them.

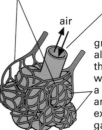

air

small air tube called a bronchiole

group of alveoli with thin, moist walls – provide a huge surface area for exchanging gases

The alveoli in the lungs have a large surface area and thin walls to allow efficient diffusion of oxygen and carbon dioxide.

B Osmosis

1 **key fact** Osmosis is the diffusion of water from a dilute solution to a more concentrated solution through a partially permeable membrane.

key fact A partially permeable membrane allows small molecules like water to pass across, but not large molecules such as starch and proteins. The cell membrane is partially permeable.

Water can move into and out of a cell if the concentration outside is different from the concentration inside the cell.

For example, in an experiment some pieces of potato are put into solutions with different sugar concentrations.

cell wall
semi-permeable cell surface membrane
vacuole
cytoplasm
solution outside the cell
nucleus
chloroplast

Here are the results of the experiment:

Potato in dilute sugar solution

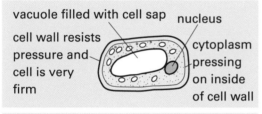

vacuole filled with cell sap
nucleus
cell wall resists pressure and cell is very firm
cytoplasm pressing on inside of cell wall

This cell has gained water by osmosis.

Potato in concentrated sugar solution

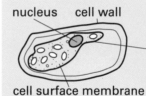

nucleus   cell wall
cytoplasm shrunken and not pressing on cell wall, so the cell is limp
cell surface membrane

This cell has lost water by osmosis.

exam tip >>

When the concentration outside a cell is the same as the concentration inside it, there is no net movement of water by osmosis.

>> practice questions

1 What is diffusion?

2 Some cut fruit is sprinkled with sugar. After a while the fruit has shrunk a little and the sugar has formed a sweet-tasting syrup. Explain, in terms of osmosis, what has happened.

Photosynthesis

- Plants make their own food by photosynthesis.

- The rate of photosynthesis may be limited by factors including carbon dioxide concentration, light intensity and temperature.

A The basics

1 Plants make their own food using a chemical reaction called photosynthesis.

2 **key fact** During photosynthesis, light energy is used to produce glucose and oxygen from carbon dioxide and water:

carbon dioxide + water $\xrightarrow{\text{light energy}}$ glucose + oxygen

sun provides light energy

CO_2 from air

CO_2

water enters roots from soil

Most photosynthesis happens in leaves. They absorb carbon dioxide from the air, and the roots absorb water from the soil.

3 Leaves contain cells with chloroplasts. These contain a green substance called chlorophyll. This absorbs the light energy needed for photosynthesis to happen.

4 Oxygen is produced as a by-product of photosynthesis, and it is released into the air from the leaves.

5 Some of the glucose produced by photosynthesis may be converted into starch. This is insoluble, and plants use it as a food store. Some of the glucose is used by the plant to release energy by respiration.

exam tip >>

Plant cells respire all the time, even when it is dark. Plants can only photosynthesise when they are in enough light.

B Factors affecting photosynthesis

(1) **key fact** The rate of photosynthesis may be limited by factors including:

- **insufficient carbon dioxide**
- **insufficient light levels**
- **low temperature.**

The rate of photosynthesis increases as the light intensity increases. But the amount of carbon dioxide available limits the maximum rate at high light levels.

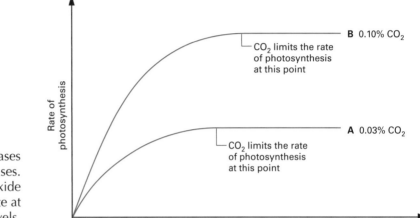

B 0.10% CO_2

CO_2 limits the rate of photosynthesis at this point

A 0.03% CO_2

CO_2 limits the rate of photosynthesis at this point

Rate of photosynthesis

Light intensity

(2) Plant growers may make an artificial environment in their greenhouses so that the plants grow quickly. They may use electric lighting, and heaters to keep the plants warm. They may also release carbon dioxide into the air to increase its concentration.

(3) Pond weed is often used to investigate the factors affecting the rate of photosynthesis. The rate of production of oxygen shows how fast photosynthesis is happening. You can find this by counting the bubbles, or by measuring the volume of oxygen that appears in a certain time.

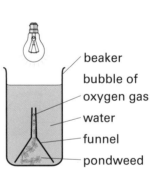

beaker
bubble of oxygen gas
water
funnel
pondweed

exam tip >>

You may be asked to analyse data showing how different factors affect the rate of photosynthesis.

>> practice questions

1 **Write the equation for photosynthesis.**

2 **State two uses of glucose by plants.**

3 **Explain why, at a particular light intensity, there may be no net release or intake of oxygen by a plant.**

Minerals

Plants need minerals including nitrate and magnesium for healthy growth.

Plants show symptoms such as poor growth if they grow in conditions where minerals are deficient.

A   Vital elements

1. Plants need certain elements to stay alive. These include hydrogen, oxygen and carbon.

2. Water from the soil provides hydrogen and oxygen. Carbon dioxide from the air provides carbon and oxygen.

3. Water and carbon dioxide are used to make food by photosynthesis. Oxygen from the air is used to release energy from food during respiration.

4. Other elements may be absorbed by plants from the soil through their roots.

B   Minerals

1. The water in soil contains dissolved mineral salts, such as nitrates and phosphates. It also contains dissolved metal ions such as magnesium and potassium.

2. Plants absorb these dissolved minerals through their roots.

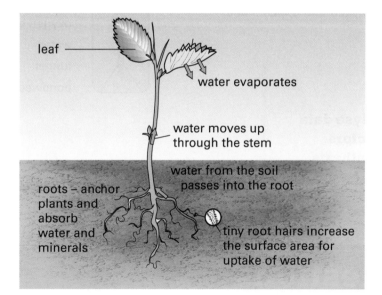

leaf

water evaporates

water moves up through the stem

water from the soil passes into the root

roots – anchor plants and absorb water and minerals

tiny root hairs increase the surface area for uptake of water

③ **key fact** The table shows why plants need two of these minerals.

mineral	why needed
nitrate	to make amino acids, which are joined together to make proteins
magnesium	to make chlorophyll

④ If plants do not get enough of these minerals, they suffer from deficiency diseases. The table shows some of the problems plants can have.

mineral deficiency	symptoms
not enough nitrate	stunted growth
not enough magnesium	yellow leaves

remember >>

Magnesium is an important component of chlorophyll, a green substance needed to absorb light energy for photosynthesis.

exam tip >>

Farmers and gardeners add fertilisers to the soil. These are chemicals that replace the minerals used by plants.

>> practice questions

1 How do plants obtain the nitrate and magnesium ions they need for healthy growth?

2 Outline why nitrate and magnesium ions are needed by plants, and the symptoms plants suffer from if these ions are deficient (in short supply).

Food chains and pyramids

 Food chains and food webs show feeding relationships in a habitat.

 Pyramids of biomass show the mass of living material at each stage in a food chain.

A Food chains

1 A food chain shows who eats what in a particular habitat.

2 **key fact** Light from the Sun is the ultimate source of energy for most living things. Green plants absorb light energy to make their own food by photosynthesis. This captured energy is stored in various substances in plant cells, such as carbohydrates and fats.

3 Other organisms in a habitat rely on the plants or other animals for their food. Omnivores eat both plants and animals. Humans are omnivores.

Type of organism	What it does
producer	Usually a green plant. It makes its own food by photosynthesis.
primary consumer	Eats plants. They are herbivores such as rabbits or horses.
secondary consumer	Eats animals. They are carnivores such as lions.

4 The arrows in a food chain point to the animal that is doing the eating. For example, cows eat grass, and humans eat cows. The food chain for these feeding relationships is:

grass → cow → human

5 Food chains are often interconnected. A food web shows these connections between food chains.

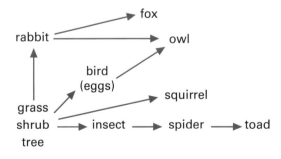

B   Pyramids of biomass

1 **key fact** The mass of living material at a particular stage in a food chain is called the biomass.

2 **key fact** A pyramid of biomass is a type of chart that shows the biomass at each stage. The width of each bar in the pyramid represents the amount of biomass. It is drawn to scale.

tertiary consumer		top carnivore
secondary consumer		carnivore
primary consumer		herbivore
producer		plant

The wider the bar, the more biomass there is at that stage.

3 **key fact** The amount of biomass decreases at each stage in a food chain.

exam tip >>

Label all the bars in your pyramid of biomass if you need to draw one. A pyramid of biomass always gets narrower as you work upwards.

>> practice questions

1 Where does the energy in food chains come from to start with?

2 What is the difference between a primary consumer and a secondary consumer?

3 What information about a food chain does a pyramid of biomass give you?

Food production

A Energy flow

1 | **key fact** | The amount of energy contained in the biomass of living things decreases at each stage in a food chain. This limits the length of the food chain. It is also an important consideration in producing food.

2 The amount of biomass and energy decreases from one stage to the next because:

- Living things produce waste materials, such as urine, faeces and carbon dioxide. These are lost to the surroundings.

- Energy is released by respiration. Most of this is lost as heat to the surroundings.

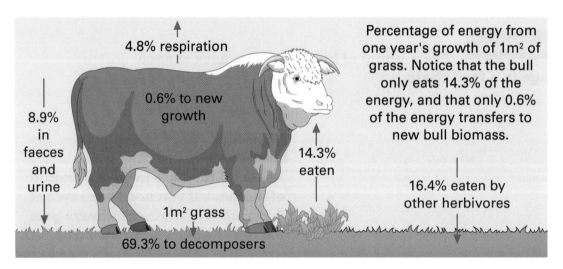

4.8% respiration

0.6% to new growth

8.9% in faeces and urine

14.3% eaten

1m² grass

69.3% to decomposers

Percentage of energy from one year's growth of 1m² of grass. Notice that the bull only eats 14.3% of the energy, and that only 0.6% of the energy transfers to new bull biomass.

16.4% eaten by other herbivores

Meat production in this bull is very inefficient due to energy losses.

B Improving efficiency

1 | **key fact** | Food chains with fewer stages are more efficient for producing food than those with more stages.

The table shows two examples of energy transfers through food chains to humans.

Food chain	Example	Approximate amount of energy available to humans (kJ per thousand hectares)
wheat → human	cereal production	10 000
grass → cattle → human	milk and meat production	milk: 4000 meat: 340

Much more energy is wasted in meat production than in cereal production.

 Heat losses from animals to the environment are particularly big in mammals (for example, cows and pigs) and birds (for example, turkeys). This is because energy from respiration is used to keep their bodies at a constant temperature.

4 Energy losses to the environment from mammals and birds can be reduced by:

- restricting their movement, for example by housing them in small pens or cages

- using heating to keep their surroundings warm.

5 There are animal welfare considerations to take into account in the way animals are housed. People may be prepared to pay more to eat food that was produced in more 'natural' ways, for example from free-range animals.

remember >>

The energy yield is greater from food chains based on plants, but the food is more difficult to digest.

exam tip >>

Be prepared to analyse information in the exam to discuss the advantages and disadvantages of different ways to improve food production.

>> practice questions

1 Outline why there is less energy at each successive stage in a food chain.

2 It has been argued that more people could be fed if everyone became vegetarians. To what extent is that view supported by the evidence in the table?

3 Explain why keeping chickens in a warm barn is more efficient in terms of energy flow than letting them run around outside.

The carbon cycle

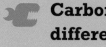 Carbon is recycled in the environment because of several different processes.

A Natural cycles

1 **key fact** In ecosystems there are processes that remove materials from the environment, and processes that return materials to it. These processes are balanced so that materials are recycled all the time.

2 Decay microorganisms digest waste materials and the remains of dead animals and plants. They can digest these faster when they are in a moist, warm environment with plenty of oxygen. Because of this decay process, substances that plants need for growth are released.

B Carbon

1 The element carbon is a part of all living things. Carbon compounds include carbon dioxide, sugars, fats, proteins and DNA.

2 Carbon is constantly recycled. All the processes involved form the carbon cycle. Carbon dioxide in the atmosphere is a huge store of carbon in the environment.

C Removing carbon dioxide from the air

1 **key fact** Green plants remove carbon dioxide from the atmosphere so they can make their own food by photosynthesis.

2 The carbon in the carbon dioxide is used by the plants to make carbohydrates such as sugars, starch and cellulose for their cell walls.

3 The carbon is also used to make fats and proteins.

remember >>

Carbon from plants becomes part of the fats and proteins in herbivores and carnivores in a food web.

① **key fact** Carbon dioxide is returned to the atmosphere because of respiration by animals, green plants, decay microorganisms and detritus feeders.

② Detritus is bits of partly broken down animal and plant material. Detritus feeders feed on this material.

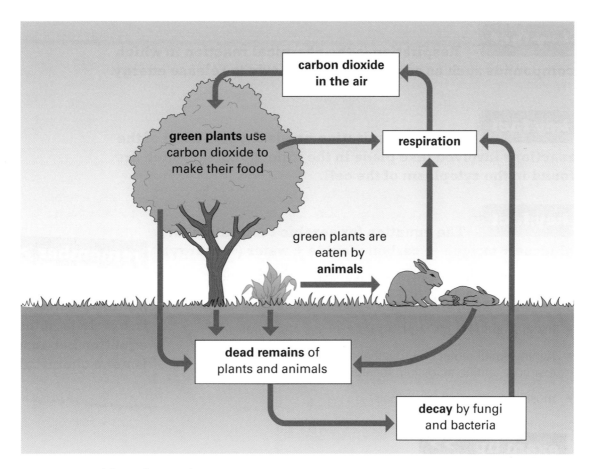

A summary of the carbon cycle.

>> practice questions

1 Name the process by which living things remove carbon dioxide from the atmosphere.

2 Name the process by which living things return carbon dioxide to the atmosphere.

3 The combustion of fossil fuels returns carbon dioxide to the atmosphere. Suggest an effect of this on the carbon cycle.

Respiration and enzymes

- Energy is released by respiration for cells to use.

- Enzymes are biological catalysts that are involved in respiration and other processes.

A Aerobic respiration

1 **key fact** Respiration is the chemical reaction in which compounds such as sugar are broken down to release energy.

2 **key fact** Aerobic respiration needs oxygen. Most of the reactions involved take place in the mitochondria, which are found in the cytoplasm of the cell.

3 **key fact** The equation for aerobic respiration is:
glucose + oxygen → carbon dioxide + water (+ energy)

4 These are some of the uses of energy released by respiration:

- maintaining the body temperature of mammals and birds

- making amino acids in plants from sugars and nitrates

- joining small molecules together to make larger ones (for example, amino acids join together to make proteins)

- muscle contraction in animals.

remember >>

Glucose is a sugar. Energy is shown in brackets in the equation because it is not a chemical.

exam tip >>

Respiration is not the same thing as breathing, which is properly called ventilation.

B Enzymes

1 **key fact** Enzymes are proteins that can increase the rate of chemical reactions. They are biological catalysts. Enzymes can catalyse reactions in living cells. These reactions include respiration, photosynthesis, and protein synthesis from amino acids.

An enzyme molecule is folded so that it has a specific shape, which is just right for other molecules to fit into. Different enzymes catalyse different reactions.

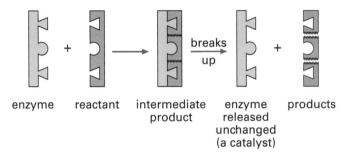

| enzyme | reactant | intermediate product | enzyme released unchanged (a catalyst) | products |

An enzyme may no longer work if its shape changes. This happens if the pH is too high or low, or the temperature is too high.

The activity of different enzymes is different at different pH values. But almost all enzymes stop working above about 60 °C. This is because they become denatured. Their shape changes so much that they cannot catalyse the reaction any more.

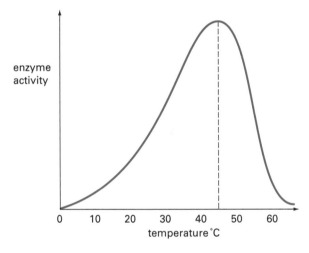

enzyme activity

temperature °C

The optimum temperature for this enzyme is 45 °C.

>> practice questions

1 What is an enzyme?

2 State three processes that are controlled by enzymes and which take place in cells.

3 Why do enzymes stop working if the temperature or pH become too high?

Digestion

 Digestive enzymes break large molecules in food into smaller molecules.

Different enzymes work in different parts of the gut.

A Secretion of enzymes

1 Some enzymes catalyse reactions outside cells. Digestive enzymes are like this.

2 **key fact** Specialised cells in the lining of the stomach and small intestine and in glands, produce enzymes. These enzymes are released from these cells into the gut.

3 **key fact** Once in the gut, the digestive enzymes catalyse the digestion or breakdown of large molecules into smaller molecules. These molecules are soluble and are small enough to pass across the gut wall into the bloodstream.

B Digestive enzymes

>> **key fact** Different enzymes digest different molecules. This table summarises the main types.

enzyme	reaction catalysed
amylase	starch → sugars
protease	proteins → amino acids
lipase	lipids → fatty acids + glycerol

Amylase is an example of a carbohydrase. Lipids are fats and oils.

C Digestive enzymes in the gut

>> **key fact** Different parts of the gut produce different enzymes. The table on the next page summarises where they are produced.

enzyme	where it is produced			
	salivary glands	stomach	pancreas	small intestine
amylase	✔		✔	✔
protease		✔	✔	✔
lipase			✔	✔

- Amylase catalyses the breakdown of starch into sugars in the mouth and small intestine.

- Proteases catalyse the breakdown of proteins into amino acids in the stomach and small intestine.

- Lipases catalyse the breakdown of fats and oils into fatty acids and glycerol in the small intestine.

D Changing pH

① **key fact** The stomach produces hydrochloric acid, which kills harmful microorganisms in the food. The proteases in the stomach work best in acidic conditions.

② **key fact** The enzymes in the small intestine work best in alkaline conditions. The liver produces a substance called bile. This is stored in the gall bladder and then released into the small intestine. It neutralises the acid produced by the stomach and makes the small intestine alkaline.

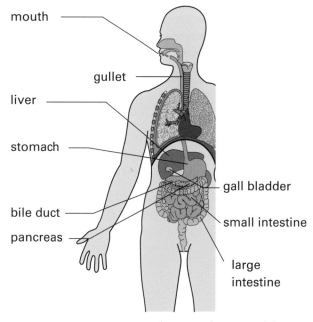

mouth
gullet
liver
stomach
bile duct
pancreas
gall bladder
small intestine
large intestine

The main features of the gut.

>> practice questions

1 **Where are carbohydrates digested?**

2 **What do lipases do?**

3 **State a difference between the proteases secreted by the stomach and those secreted by the pancreas and small intestine.**

Uses of microorganisms and enzymes

Enzymes released from microorganisms are used in detergents, and in the food industry.

A The basics

① key fact Some microorganisms produce enzymes that are released from the cells. For example, certain yeasts (single-celled fungi) and bacteria do this.

② The microorganisms can be grown in huge numbers in containers called fermenters. The enzymes they produce are separated and purified for use in the home and industry.

B Enzymes in the home

① The stains on dirty clothes contain fats, oils and proteins. These substances are difficult to remove because they stick to the fibres in the clothing and they do not dissolve in water.

② Biological detergents contain lipases to digest the fat and oils in stains. The fatty acids and glycerol produced in the reaction are soluble in water, so they are more easily removed during washing.

③ Biological detergents also contain proteases to digest the proteins in stains. The amino acids produced in the reaction are soluble in water, so they are more easily removed during washing.

④ Clothes cannot be washed in water above 40 °C with biological detergents. The lipases and proteases are denatured at high temperatures and so stop working in hot water.

1. The digestive system in babies is not fully developed. The proteins in some baby foods are pre-digested using proteases.

2. Carbohydrase converts starch syrup into sugar syrup. Starch syrup is not sweet but is relatively inexpensive, but sugar syrup is sweet and more expensive. It is often used in sports drinks. The use of carbohydrase enzymes makes a valuable product from a cheap raw material.

3. Isomerase converts glucose syrup into fructose syrup. Fructose syrup is sweeter than the same mass of glucose syrup. It is used in slimming foods because less of it is needed to sweeten the food.

4. Enzymes are also used in some industrial chemical reactions. They allow the reactions to work at high rates at normal temperatures and pressures. This reduces the energy needed and the cost of the equipment.

> **exam tip >>**
>
> **The production of wine and beer relies on fermentation. This process needs the enzymes contained in yeast. You will study fermentation if you are doing GCSE biology.**

>> practice questions

1 Explain how biological detergents work.

2 Clothes can be washed at relatively low temperatures using biological detergents. Suggest why this may reduce the impact on the environment of washing clothes.

3 State the benefits of using carbohydrase and isomerase in food production.

Maintaining internal conditions

The internal conditions of the body are controlled so that they stay roughly the same.

A Internal conditions

1 Cells work best if the conditions stay the same. Waste products would increase in concentration if they were not removed. Nutrients and other chemicals needed by the cell would decrease in concentration if they were not replaced.

2 **key fact** The conditions inside the body that are controlled include the:

- **ion content**
- **water content**
- **temperature**
- **blood sugar level.**

B Ions and water

1 The ion and water content of the body influence the concentration of dissolved substances:

- If the solution outside a cell is too concentrated, water leaves the cell.
- If the solution outside a cell is too dilute, water enters the cell.

In both cases the cell can be damaged.

2 **key fact** The body gains ions and water through food and drink.

3 **key fact** The body loses ions and water through urine and sweat. We sweat more when it is hot, so we need to take in more water and ions by eating and drinking.

remember >>

Water is also lost from the lungs when we breathe out.

C Waste products

 key fact Carbon dioxide is a waste product of respiration. It is lost from the lungs when we breathe out.

key fact Excess amino acids are broken down by the liver. Urea is produced by this process. Urea is removed from the bloodstream by the kidneys. The urine produced by the kidneys is stored in the bladder, before leaving the body.

D Body temperature

key fact A certain part of the brain is called the thermoregulatory centre. It controls and monitors body temperature. It has receptors that are sensitive to the temperature of the blood flowing in the brain.

key fact Receptors in the skin transmit information about skin temperature to the thermoregulatory centre as nerve impulses.

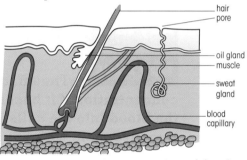

A cross-section of the skin.

Higher Tier

key fact Sweat glands and blood capillaries in the skin are involved in adjusting the body temperature.

	skin capillaries	sweat glands	muscles
too hot	Blood vessels **dilate** (widen) so more blood goes through the skin and more heat is lost.	More sweat is released, which evaporates and cools the skin.	No shivering.
too cold	Blood vessels **constrict** (narrow) so less blood goes through the skin and less heat is lost.	Less sweat is released.	Muscles **shiver** so they respire more as they move, which releases heat.

>> practice questions

1 State four internal conditions of the body that are controlled by the body.

2 Where does urea come from and how it is removed from the body?

3 What are the functions of the thermoregulatory centre?

Diabetes

- Diabetes is a condition where blood glucose concentration is not controlled sufficiently by hormones.

- Most diabetes is controlled by injecting insulin.

A Blood glucose concentration

1. Glucose is a simple sugar. It is transported round the body in the bloodstream. This is so that the body's cells can get the glucose they need for respiration.

2. The blood glucose concentration is important. If it is too high, water leaves the cells by osmosis and the cells become damaged, and in addition, it could lead to coma and even death.

B The pancreas and insulin

1. **key fact** The pancreas produces a hormone called insulin.

2. **key fact** The effect of insulin is to reduce the blood glucose concentration if it becomes too high. It allows glucose to move from the blood into the cells, and it causes the liver to convert glucose into glycogen, which is stored.

blood glucose concentration too high

⇩

pancreas releases more insulin

⇩

cells take up more glucose

⇩

liver converts glucose into glycogen for storage

⇩

blood glucose concentration lowered

exam tip >>

Insulin was first isolated in 1921 by two doctors, Frederick Banting and Charles Best. You may be given data from their experiments to evaluate.

24

C **Diabetes**

1 **key fact** Diabetes is a disorder where the blood glucose concentration can increase to a level that is fatal.

2 The symptoms of diabetes include feeling unusually thirsty, excreting sugar in the urine, and lack of energy.

3 In type 1 diabetes, the pancreas does not produce enough insulin. It can be treated by:

- careful control of the diet and

- injections of insulin.

Diabetics can monitor their blood glucose concentration using a glucometer. A drop of blood from a pin prick is used to get an instant read-out.

exam tip >>

In GCSE Science you discovered that nowadays most insulin for treating diabetes is made from genetically modified bacteria.

>> practice questions

1 Where is insulin produced?

2 What is diabetes?

3 Explain how an injection of insulin affects blood glucose concentration.

DNA

A   DNA

1 DNA is deoxyribose nucleic acid. It is a large, complex molecule.

2 **key fact** A gene is a small section of DNA. Each gene codes for a certain combination of amino acids that makes a particular protein.

3 **key fact** Some characteristics are controlled by a single gene but others are controlled by several genes.

4 **key fact** Everyone's DNA is unique. Only identical twins share the same DNA.

5 DNA fingerprinting, also called genetic fingerprinting, relies on this. Enzymes are used to cut a sample of DNA into short sections, which are then separated by a type of chromatography called gel electrophoresis. A unique pattern of bands is formed, rather like a bar code.

B   Chromosomes

1 Chromosomes are made from large molecules of DNA. They are found in the nucleus of the cell. A single chromosome may contain several hundred genes, or more.

2 A small section of DNA that carries the code for a particular protein is called a gene.

3 Chromosomes are always in pairs in body cells such as muscle or skin cells. Each human body cell has 23 pairs of chromosomes in its nucleus.

Two chromosomes are the sex chromosomes. They determine the sex of an individual:

- males have two different sex chromosomes, XY
- females have the same two sex chromosomes, XX.

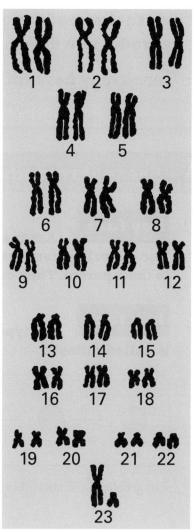

The 23 pairs of chromosomes from a human male. His XY sex chromosomes are numbered 23.

>> practice questions

1 **What is DNA and where is it found?**

2 **How many pairs of chromosomes are found in human body cells?**

3 **What is the difference between the sex chromosomes in men and women?**

Cell division

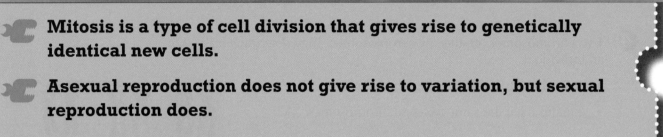
A Asexual reproduction

1 **key fact** Asexual reproduction involves just one parent. The offspring from asexual reproduction are genetically identical to their parent. They contain the same genes.

2 **key fact** The type of cell division in asexual reproduction is called mitosis.

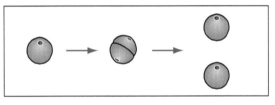

Two genetically identical new cells are formed by mitosis.

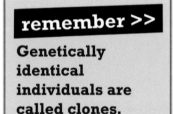

remember >>

Genetically identical individuals are called clones.

B Sexual reproduction

1 **key fact** Sexual reproduction involves two parents. Each parent produces sex cells or gametes. The male gamete is the sperm, and the female gamete is the egg.

2 Gametes form by a different type of cell division than mitosis. The gametes are genetically different from each other. Gametes contain half the number of chromosomes that body cells do. For example, human gametes contain only 23 chromosomes.

3 **key fact** Fertilisation is the fusion or joining of a male gamete with a female gamete to form a new cell.

④ key fact The new cell or embryo contains the same number of chromosomes as its parents' body cells. For example, human embryos contain 23 pairs of chromosomes.

D Variation

① key fact Each gene may have different forms called alleles:

• body cells contain pairs of alleles, one on each chromosome in a pair of chromosomes;

• gametes contain just one allele from each pair of alleles on its parents' chromosomes.

② key fact When an embryo forms, half of each pair of alleles comes from one parent, and half of each pair comes from the other parent. This gives rise to genetic variation. The offspring gets half of its genetic information from each parent.

③ Once formed, the embryo cell develops into a new individual by repeatedly dividing by mitosis.

>> practice questions

1 Why are the offspring from asexual reproduction genetically identical?

2 Why do the offspring from sexual reproduction show genetic variation?

3 What is an allele?

Inheritance

- Genes are responsible for the inheritance of different characteristics.

- Alleles can be dominant or recessive.

- Genetic diagrams let us predict and understand the results of different crosses.

A The basics

1 Individual genes are inherited separately from one another. They are responsible for the inheritance of different characteristics.

2 Some characteristics, such as eye colour, are controlled by a single gene. Others, such as height, are controlled by many genes.

3 If only one or a few genes are involved, you can predict patterns of inheritance using genetic diagrams.

B Mendel's work

1 Gregor Mendel was a German monk who carried out experiments on pea plants in the 19th century. He carried out many crosses or breeding experiments between pea plants. He studied different characteristics including flower colour.

2 Mendel started with plants that had produced only red flowers or white flowers for several generations. These plants would only carry one allele for flower colour.

3 When he bred or crossed red flowered plants with white flowered plants, the seeds only produced red flowered plants.

4 When he crossed these red flowered plants, some of the seeds produced white flowered plants again.

5 The reasons for his results can be explained using genetic diagrams.

exam tip >>

You may be asked to evaluate information about Mendel's work, and you should be able to interpret genetic diagrams.

C Dominant and recessive alleles

① key fact A dominant allele controls the development of a particular characteristic even when it is found on only one of the chromosomes in a pair.

② key fact A recessive allele only controls the development of a particular characteristic if the dominant allele is not present. Two copies of a recessive allele are needed to control the development of the characteristic.

D Genetic diagrams

In genetic diagrams, a dominant allele is shown as a capital letter and a recessive allele is shown as a lowercase letter. For example, in Mendel's pea plant experiment we could show the dominant allele for flower colour as F, and the recessive allele for flower colour as f.

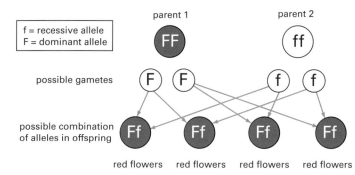

A genetic diagram for Mendel's first cross.

Red flowered parent plants contain two copies of the dominant allele for flower colour. White flowered parent plants contain two copies of the recessive allele for flower colour. All the offspring contain one copy of each allele, but the dominant allele gives rise to all red flowers.

A genetic diagram for Mendel's second cross.

The red flowered parent plants from the first generation contain one copy of each allele for flower colour. The offspring with one or two copies of the dominant allele produce red flowers. Offspring with two copies of the recessive allele produce white flowers.

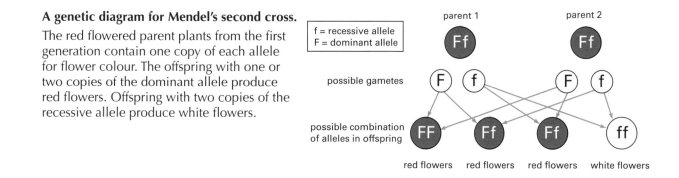

>> practice questions

1 In the genetic diagrams above, why do plants with the alleles FF and Ff produce red flowers but plants with the alleles ff produce white flowers?

Inherited disorders

> Huntington's disease is a genetic disorder caused by a dominant allele.

> Cystic fibrosis is a genetic disorder caused by a recessive allele.

A The basics

1. Genetic disorders are inherited. Huntington's disease and cystic fibrosis are two genetic disorders.

2. The chance of two parents producing a child with a genetic disorder can be predicted using a genetic diagram. Embryos can be tested to see if they carry the allele responsible for a certain disorder. This is called embryo screening. The parents can make a decision about the future of the embryo before it develops into a baby.

> **exam tip >>**
>
> There are social and ethical issues surrounding embryo screening. In the examination, you may be given information about embryo screening to discuss and evaluate.

B Huntington's disease

1. **key fact** Huntington's disease is a disorder of the nervous system. It is caused by a dominant allele. The allele causes the production of a protein that damages the brain, which affects memory and body movements.

2. **key fact** Huntington's disease can be passed on by just one parent, because it is caused by a dominant allele. Only one copy of the allele is needed.

In this genetic diagram, the mother has Huntington's disease, even though she only has one copy of the Huntington's allele. The father does not have the allele.

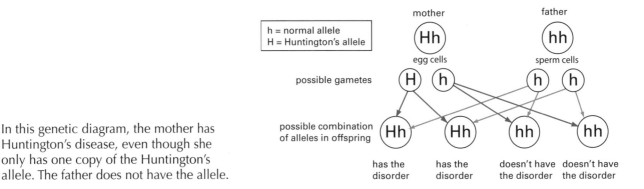

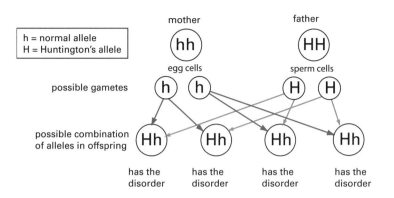

h = normal allele
H = Huntington's allele

mother

hh

father

HH

egg cells

sperm cells

possible gametes

h   h   H   H

possible combination of alleles in offspring

Hh   Hh   Hh   Hh

has the disorder   has the disorder   has the disorder   has the disorder

In this genetic diagram, the mother does not have Huntington's disease. The father does, and he carries two copies of the allele. All the offspring would have Huntington's disease.

C Cystic fibrosis

1 key fact Cystic fibrosis causes thick, sticky mucus in the lungs and gut. It is caused by a recessive allele.

2 key fact Cystic fibrosis must be passed on by both parents, because two copies of the fault allele are needed.

3 key fact A carrier of cystic fibrosis has one copy of the faulty allele, so they do not have the disorder. But they can pass the disorder on to their children if the other parent is also a carrier.

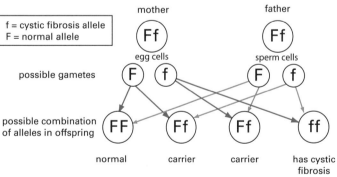

f = cystic fibrosis allele
F = normal allele

mother

Ff

father

Ff

egg cells

sperm cells

possible gametes

F   f   F   f

possible combination of alleles in offspring

FF   Ff   Ff   ff

normal   carrier   carrier   has cystic fibrosis

In this genetic diagram, both parents are carriers of the cystic fibrosis allele. One possible offspring would be normal. Two would be carriers themselves, and the remaining offspring would have cystic fibrosis.

>> practice questions

1 Explain why Huntington's disease can be passed on by just one parent who has the disorder.

2 What is a carrier?

3 Explain why cystic fibrosis must be passed on by both parents.

Gametes are formed by a type of cell division called meiosis.

A Mitosis

1 Chromosomes are found in pairs in body cells. So body cells have two sets of genetic information.

2 Body cells divide by mitosis. This provides extra cells for growth or repair. The new cells are genetically identical to the original cell.

B Meiosis

1 The reproductive organs in humans are:

- the testes in males
- the ovaries in females.

remember >>

The singular of testes is testis.

2 Cells in the reproductive organs divide to form gametes: sperm cells in males and egg cells in females.

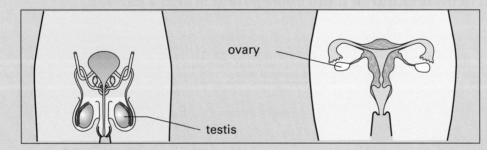

The male and female reproductive systems, with the locations of the testes and ovaries indicated.

3 Gametes are formed by a type of cell division called meiosis. In meiosis:

- the chromosomes are copied
- the cell divides twice
- four gametes are formed, each with a single set of chromosomes.

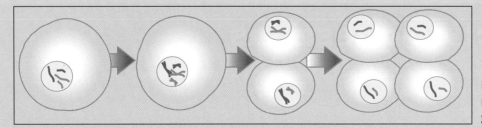

Four genetically different new cells are formed by gamete production.

C Predicting outcomes of crosses

A Punnett square is another type of genetic diagram. It can be used to show genetic crosses very quickly. The alleles present in the egg cells are shown at the top, and the alleles present in the sperm cells are shown at the side.

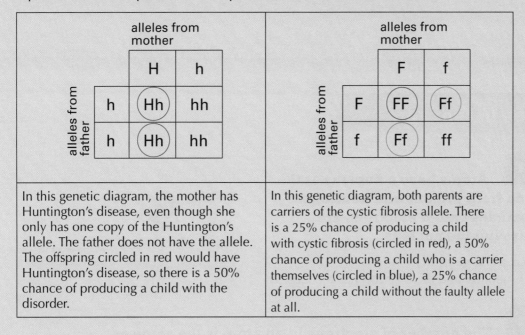

In this genetic diagram, the mother has Huntington's disease, even though she only has one copy of the Huntington's allele. The father does not have the allele. The offspring circled in red would have Huntington's disease, so there is a 50% chance of producing a child with the disorder.

In this genetic diagram, both parents are carriers of the cystic fibrosis allele. There is a 25% chance of producing a child with cystic fibrosis (circled in red), a 50% chance of producing a child who is a carrier themselves (circled in blue), a 25% chance of producing a child without the faulty allele at all.

exam tip >>

You should be able to predict and explain the outcome of any cross when you are given information about the dominant and recessive alleles.

>> practice questions

1 **Outline what happens in meiosis.**

2 **The allele for blue eyes in humans is recessive (b), and the allele for brown eyes is dominant (B).**

   a) **Predict the outcomes of this cross involving brown-eyed parents: Bb × Bb.**

   b) **What proportion of offspring will have blue eyes?**

Atomic structure

- Atoms contain protons, neutrons and electrons.

- All the atoms of a particular element have the same number of protons.

- Electrons are arranged around the nucleus in energy levels or shells.

A Particles in the atom

1 key fact Atoms have a nucleus at the centre made from protons and neutrons. Smaller particles called electrons are arranged around the nucleus.

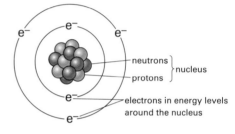

2 The proton has a charge of +1.
The neutron is neutral, and the electron has a charge of –1.

3 key fact The number of electrons in an atom is the same as the number of protons. This means that atoms are neutral overall.

B Atomic number and the periodic table

1 key fact The number of protons in the nucleus of an atom is its atomic number (also called the proton number).

2 key fact Atoms with the same atomic number belong to the same element. Atoms with different atomic numbers belong to different elements.

3 key fact Elements in the periodic table are arranged in order of increasing atomic number. The atomic number is shown to the bottom left of each symbol.

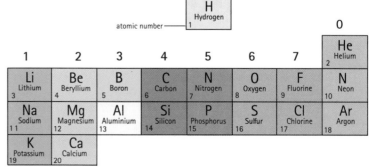

A short form of the periodic table showing the first 20 elements.

Electronic structure

1 **key fact** Electrons occupy energy levels, starting with the lowest available energy level. The way in which an atom's electrons are arranged in the energy levels is called its electronic structure.

2 The periodic table can be used to work out the electronic structure of any one of the first twenty elements. You count along to the element from hydrogen, putting in a comma if you have to go down to another period (row in the periodic table).

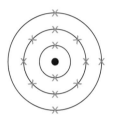

The electronic structure of aluminium is 2,8,3.
You can draw the atom like this.
Each energy level is a circle, and each electron is a cross.

3 There are three checks:

- the last number in an electronic structure is the same as the group number

- the number of numbers is the same as the period number

- the total number of electrons is the same as the atomic number.

exam tip >>

You will be given a copy of the periodic table in the exam. If you prefer, you can answer questions in terms of shells instead of energy levels.

>> practice questions

1 Write the electronic structures of the first twenty elements, starting at H 1 and finishing with Ca 2,8,8,2.

Ions and ionic bonds

- Compounds consist of atoms of two or more elements chemically combined.

- Atoms lose or gain electrons to form charged particles called ions.

- Ionic compounds contain oppositely charged ions held together by strong forces of attraction called ionic bonds.

A Ions

An ion is a charged particle formed when an atom loses or gains electrons. It has the same electronic structure as one of the noble gases in Group 0. The charge on an ion is shown at the top right of its symbol.

B Positive ions

1 Hydrogen atoms and metal atoms lose electrons to form positive ions.

2 **key fact** A metal atom loses the electrons from its highest occupied energy level. The charge on the ion is the same as the group number. Sodium in Group 1 forms Na+ ions; magnesium in Group 2 forms Mg^{2+} ions; and calcium in Group 2 forms Ca^{2+} ions.

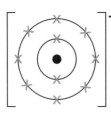 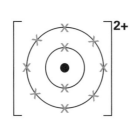 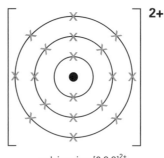

sodium ion $[2,8]^+$ magnesium ion $[2,8]^{2+}$ calcium ion $[2,8,8]^{2+}$

C Negative ions

1 Non-metal atoms gain electrons to form negative ions.

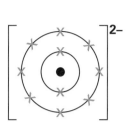

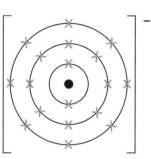

key fact A non-metal atom gains enough electrons to complete its highest occupied energy level.

The number of charges on the ion is eight minus the group number. For example, oxygen in Group 6 forms O^{2-} ions. Chlorine in Group 7 forms Cl^- ions.

oxide ion $[2,8]^{2-}$ chloride ion $[2,8,8]^-$

D Ionic bonds

In a chemical reaction between a metal and a non-metal, electrons are transferred from the metal atoms to the non-metal atoms. The new substance formed is called an ionic compound.

key fact Ionic compounds contain positive ions and negative ions held together in a giant structure. There is a strong force of attraction between these oppositely charged ions. It acts in all directions and is called ionic bonding.

Dot and cross diagrams show how electrons are transferred during a chemical reaction. The electrons in one atom are shown as dots, and the electrons in the other atom are shown as crosses.

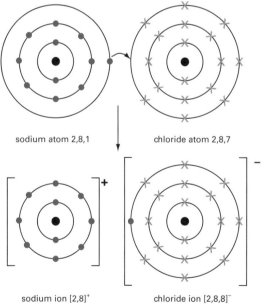

sodium atom 2,8,1   chloride atom 2,8,7

A dot and cross diagram to show the formation of sodium chloride.

sodium ion $[2,8]^+$ chloride ion $[2,8,8]^-$

>> practice questions

1 **Draw dot and cross diagrams to show the formation of:**

a) **magnesium oxide, and**

b) **calcium chloride.**

Simple molecules

A Covalent bonds

 Covalent bonds occur between non-metal atoms.

key fact A covalent bond forms when a pair of electrons is shared between two atoms. These bonds are very strong.

Only electrons in the highest occupied energy level can be shared. Electrons are shared so that each atom can complete its highest occupied energy level.

B Representing covalent bonds

 Covalent bonds can be shown using dot and cross diagrams. Only the highest energy levels are drawn. They are shown overlapping, with the shared pair of electrons in the overlapping area.

Covalent bonds can also be shown as straight lines in displayed formulae.

A hydrogen atom can form one covalent bond. The atoms in Group 0 do not form covalent bonds under most conditions. For other non-metal atoms, the number of bonds formed is eight minus the group number.

For example, methane CH_4 is a compound of carbon and hydrogen. Carbon is in Group 4, so it forms 8 – 4 = 4 covalent bonds. Hydrogen atoms can each form one covalent bond.

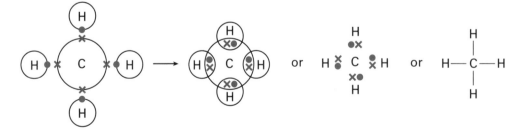

Three ways to show the covalent bonding in methane.

C Other simple molecules

A simple molecule consists of only a few atoms joined together by covalent bonds. Elements such as hydrogen, chlorine and oxygen form simple molecules. Compounds such as methane, hydrogen chloride, ammonia and water form simple molecules.

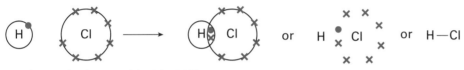

Bonding in hydrogen, H_2.

Bonding in chlorine Cl_2. Chlorine is in Group 7, so it forms $8 - 7 = 1$ bond.

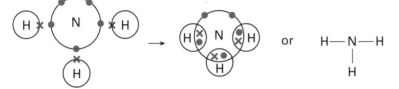

Bonding in hydrogen chloride, HCl.

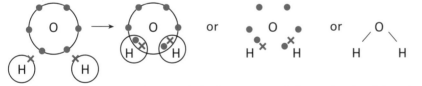

Bonding in ammonia, NH_3. Nitrogen is in Group 5, so it forms $8 - 5 = 3$ bonds.

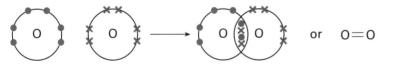

Bonding in water, H_2O. Oxygen is in Group 6, so it forms $8 - 6 = 2$ bonds.

Bonding in oxygen, O_2.

remember >>

One shared pair of electrons forms a single bond. Two shared pairs of electrons form a double bond. There is a double bond in an oxygen molecule.

>> practice questions

1 **What is a covalent bond?**

2 **Draw a dot and cross diagram to represent the bonds in nitrogen, N_2.**

Structure and properties of materials

- Substances with giant covalent structures or giant ionic lattices have high melting points. Substances with simple molecules have low melting points.

- Metals have layers of atoms that slide over each other easily.

- Nanoparticles have different properties from the same substances in bulk.

A Giant covalent structures

1 **key fact** Some covalently bonded substances have giant covalent structures in which all the atoms are linked to other atoms. They are also called macromolecules.

2 **key fact** Diamond, graphite and silicon dioxide (silica) have giant covalent structures. They have very high melting points because they contain very many strong covalent bonds.

exam tip >>

You will be expected to recognise, and describe, the structure and bonding in diamond and graphite, but not to draw them.

diamond

Diamond structure

graphite

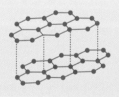

Graphite structure

Each carbon atom covalently bonded to four other carbon atoms.	Each carbon atom covalently bonded to three other carbon atoms.
Regular lattice of carbon atoms with strong covalent bonds makes diamond very hard.	Layers of carbon atoms with weak forces between them. Layers can slide over each other, so graphite is soft and slippery.

B Simple molecules

1 **key fact** Substances that consist of simple molecules have relatively low melting points and boiling points. They are usually gases such as oxygen and carbon dioxide, or liquids such as water and ethanol.

key fact Simple molecules do not have an overall electric charge, so they do not conduct electricity.

C Ionic compounds

key fact Ionic compounds have regular structures called giant ionic lattices. Oppositely charged ions are attracted to each other by strong electrostatic forces called ionic bonds. Giant ionic lattices have high melting points and boiling points.

key fact Ionic compounds do not conduct electricity when they are solid. But they do conduct electricity when they are dissolved in water, or melted. This is because their ions are free to move.

D Metals

>> key fact Metals can be bent and shaped without shattering. This is because they have layers of metal atoms that can slide over each other.

E Nanoparticles

key fact One nanometre is one millionth of a millimetre. Nanoparticles are between 1 nm and 100 nm in size, and each contain just a few hundred atoms.

key fact Nanoparticles have a high surface area to volume ratio. They have different properties from the same substance in larger pieces.

>> practice questions

1 Why is graphite soft and slippery but diamond hard?

2 Why does sodium chloride conduct electricity when it is dissolved in water but not when it is solid?

3 What types of structures lead to high melting points?

Structure and bonding – *Higher Tier*

- Substances with giant covalent structures or giant ionic lattices have high melting points. Substances with simple molecules have low melting points.

- Metals have layers of atoms that slide over each other easily.

- Nanoparticles have different properties from the same substances in bulk.

A Metallic bonding

1 **key fact** Metals have giant structures. Their atoms are arranged in a regular way.

2 **key fact** The electrons in the highest occupied energy level leave the atoms. They become free to move through the entire structure. These free electrons are described as being delocalised because they are not associated with any particular metal atom.

3 **key fact** In a metal, the atoms actually exist as positive ions because the electrons from the highest occupied energy level have become delocalised.

4 **key fact** There are strong electrostatic forces of attraction between the positively charged metal ions and the negatively charged delocalised electrons. These forces are called metallic bonds.

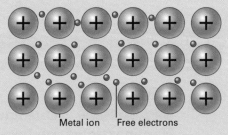

Metal ion   Free electrons

B Conduction in metals

1 Metals are good conductors of electricity and heat.

2 **key fact** Metals conduct electricity because their delocalised electrons are free to move and carry charge through the metal.

3 **key fact** Metals conduct heat because of their delocalised electrons. The delocalised electrons gain kinetic energy when the metal is heated. They transfer this energy throughout the metal as they move.

C Electrical conduction by graphite

1 **key fact** Graphite has a giant covalent structure but it conducts electricity.

2 **key fact** Carbon is in Group 4, so it can form 8 – 4 = 4 covalent bonds. In graphite, each carbon atom forms only three covalent bonds. One electron from each carbon atom becomes delocalised.

3 **key fact** The delocalised electrons are free to move. They allow graphite to conduct electricity and heat.

D Intermolecular forces

1 **key fact** The atoms in a simple molecule are joined together by strong covalent bonds. But there are only weak forces attracting simple molecules towards each other.

2 **key fact** The weak forces between simple molecules are called intermolecular forces.

> **exam tip >>**
> You may be asked to suggest the type of structure that a particular substance has, if you are told its properties.

3 **key fact** When a substance consisting of simple molecules melts or boils, the weak intermolecular forces are overcome, not the strong covalent bonds. This is why these substances have relatively low melting and boiling points.

type of structure	simple molecules	giant covalent	giant ionic	metallic
bonding	covalent	covalent	ionic	metallic
melting and boiling points	low	high	high	high
conduct electricity when solid?	no	no (except graphite)	no (yes when liquid or dissolved in water)	yes

>> practice questions

1 Describe the structure and bonding in metals.

2 Why do metals and graphite conduct electricity?

3 Why do substances with simple molecules have relatively low melting and boiling points?

Relative atomic mass

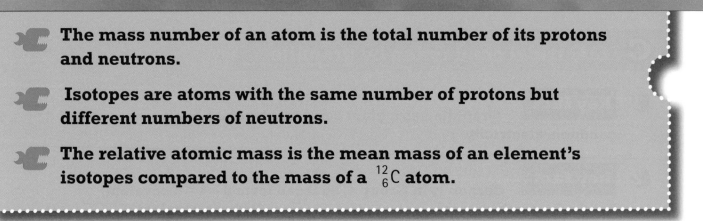

- The mass number of an atom is the total number of its protons and neutrons.
- Isotopes are atoms with the same number of protons but different numbers of neutrons.
- The relative atomic mass is the mean mass of an element's isotopes compared to the mass of a $^{12}_{6}C$ atom.

A Mass number

1 key fact The protons, neutrons and electrons in an atom have mass.

2 key fact The total number of protons and neutrons in the nucleus of an atom is called the mass number.

3 The table shows the relative masses of protons, neutrons and electrons.

particle	relative mass
proton	1
neutron	1
electron	very small

4 key fact The full chemical symbol for an atom shows its mass number and its atomic number (proton number). For example, for sodium:

mass number ⟶ 23
atomic number ⟶ 11 Na ⟵ symbol

5 The number of neutrons = mass number – atomic number.

For example, for $^{23}_{11}Na$, there are 23 – 11 = 12 neutrons.

There are 11 protons (and so 11 electrons).

remember >>

Protons and neutrons are found in the nucleus. This means that most of the mass of an atom is found there.

B Isotopes

1 key fact Isotopes are atoms of an element that have different numbers of neutrons. They have the same atomic number but different mass numbers.

2 For example, $^{12}_{6}C$ and $^{14}_{6}C$ are isotopes of the same element. $^{40}_{18}Ar$ and $^{40}_{19}K$ are isotopes of different elements (their atomic numbers are different, even though they have the same mass number).

C Relative atomic mass

1 key fact The relative atomic mass of an element is the mean mass of its isotopes compared to the mass of the $^{12}_{6}C$ atom. It has the symbol A_r.

2 The A_r of $^{12}_{6}C$ is 12 exactly. Atoms of $^{48}_{22}Ti$ are four times heavier.

3 Many elements have A_r values that are whole numbers. But chlorine has two isotopes (75% are $^{35}_{17}Cl$ and 25% are $^{37}_{17}Cl$). Its A_r is 35.5 because:

$$A_r = \frac{(75 \times 35) + (25 \times 37)}{75 + 25} = \frac{2625 + 925}{100} = \frac{3550}{100} = 35.5$$

D Relative formula mass

1 key fact The relative formula mass of a compound is found by adding together the A_r values of all the atoms in its formula. It has the symbol M_r. It is also called the relative molecular mass.

2 For example, the M_r of $Mg(OH)_2 = 24 + 16 + 16 + 1 + 1 = 58$.

exam tip >>

You will be given a copy of the periodic table in the exam. It shows the relative atomic mass of each element.

>> practice questions

1 What are isotopes of an element?

2 How many times heavier than an atom of $^{12}_{6}C$ is an atom of $^{84}_{36}Kr$?

3 $A_r(Ca) = 40$, $A_r(O) = 16$, $A_r(H) = 1$ and $A_r(S) = 23$.

Work out the relative formula mass of:

a) CaO   b) H_2S c) $CaSO_4$ d) $Ca(OH)_2$

Formulae and percentage composition

- The formula of an ionic compound can be worked out from the formulae of its ions.

- The percentage composition by mass of an element in a compound can be worked out using the formula and M_r of the compound, and the A_r of the element.

- The relative formula mass of a compound in grams is one mole of the compound.

A Formulae of ionic compounds

1 You can work out its formula if you know the formulae of the ions it contains.

positive ions		negative ions	
name	**formula**	**name**	**formula**
aluminium	Al^{3+}	bromide	Br^-
ammonium	NH_4^+	carbonate	CO_3^{2-}
barium	Ba^{2+}	chloride	Cl^-
calcium	Ca^{2+}	fluoride	F^-
copper(II)	Cu^{2+}	hydroxide	OH^-
hydrogen	H^+	iodide	I^-
iron(II)	Fe^{2+}	nitrate	NO_3^-
iron(III)	Fe^{3+}	oxide	O^{2-}
lead	Pb^{2+}	sulfate	SO_4^{2-}
lithium	Li^+	sulfide	S^{2-}
magnesium	Mg^{2+}		
potassium	K^+		
silver	Ag^+		
sodium	Na^+		
zinc	Zn^{2+}		

The formulae of some common positive ions and negative ions.

2 **key fact** The total number of positive charges in an ionic compound is equal to the total number of negative charges. Here are three examples:

- The formula of sodium chloride is NaCl. The single positive charge on the Na^+ ion balances the single negative charge on the Cl^- ion.

- The formula of iron(III) oxide is Fe_2O_3.
  There are $2 \times 3 = 6$ positive charges on the Fe^{3+} ions.
  These balance the $3 \times 2 = 6$ negative charges on the O^{2-} ions.

- The formula of ammonium carbonate is $(NH_4)_2CO_3$. There are $2 \times 1 = 2$ positive charges on the NH_4^+ ions. These balance the 2 negative charges on the CO_3^{2-} ions. Note that NH_4 goes inside brackets because it contains two elements and more than one of it is needed.

B Percentage composition

1 key fact The percentage by mass of an element in a compound can be calculated. To do this, you need to know:

- the formula of the compound
- the relative atomic mass of the element involved
- the relative formula mass of the compound.

2 key fact % mass of element X =

$$\frac{\text{number of atoms of X in formula} \times A_r \text{ of X}}{M_r \text{ of compound}} \times 100$$

3 For example, what is the percentage by mass of O in SO_2?

A_r of S = 32 and A_r of O = 16

M_r of SO_2 = 32 + 16 + 16 = 64

There are two atoms of O in SO_2

% mass of O = $\frac{2 \times 16}{64} \times 100 = \frac{32}{64} \times 100 = 50\%$

C The mole

1 key fact The relative formula mass of a substance written in grams is one mole of that substance.

2 The relative formula mass of sulfur dioxide is 64. So one mole of sulfur dioxide molecules has a mass of 64 g.

>> practice questions

1 What is the formula of:

a) calcium oxide,

b) iron(III) sulfide,

c) aluminium hydroxide?

2 The M_r of NaOH is 40. What is the percentage by mass of O in NaOH?

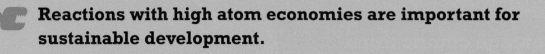

Atom economy and balancing equations

- Reactions with high atom economies are important for sustainable development.

- The yield of a particular product in a chemical reaction may be less than expected.

- Chemical equations can be balanced by adjusting the numbers of atoms and molecules they contain, without changing any formulae.

A Atom economy

1. **key fact** The atom economy of a chemical reaction is a measure of how much of the starting materials becomes useful products. It is also called atom utilisation.

2. **key fact** Reactions with high atom economies produce less waste than reactions with low atom economies. They may be cheaper to run and will use less of the Earth's resources.

3. Sustainable development means making sure we can meet our own needs without harming the ability of future generations to meet their needs. Reactions with high atom economies are important for sustainable development.

B Yield in reactions

1. **key fact** The mass of a product made in a chemical reaction is called its yield. It is possible to calculate the expected yield. The actual yield may be less than expected.

2. The actual yield may be less than the expected yield because:

   • some of the product may be lost when it is separated from the reaction mixture

   • if the reaction is reversible it may not go to completion

   • other reactions may happen in addition to the expected reaction.

exam tip >>

You must be able to calculate atom economy and percentage yield if you are taking the Higher Tier.

C Balancing equations

 1 This is a word equation: magnesium + oxygen → magnesium oxide. It shows you the reactants and products in the reaction, but it does not tell you about the quantities involved.

2 For quantitative information, the word equation must be changed into a balanced chemical equation.

3 The first step is to change the words into symbols and formulae.
In the example, this becomes: $Mg + O_2 → MgO$

- Elements have symbols, so magnesium becomes Mg.

- Apart from the gases in Group 0, all the gaseous elements are diatomic (two atoms combined) so oxygen becomes O_2.

- Compounds have formulae, so magnesium oxide becomes MgO.

4 The second step is to check the equation to see if it is balanced. The number of atoms of each element on the left of the arrow must be the same as the number on the right of it.
In the equation $Mg + O_2 → MgO$:

- there is one Mg on each side, so Mg is balanced

- but there are two Os on the left and only one on the right, so O is not balanced.

5 The next step is to balance the equation. Formulae must never be changed. Instead, you can put numbers in front of a formula to adjust the number of atoms or molecules present.

- MgO is multiplied by 2 to give: $Mg + O_2 → 2MgO$

- O now balances, as there are two Os on each side, but Mg no longer balances (there is one on the left and two on the right).

- Mg is multiplied by 2 to give: $2Mg + O_2 → 2MgO$

This is balanced because there are two Mg and two O on both sides.

6 The final step is to add state symbols. (s) means solid; (l) means liquid; (g) means gas; and (aq) means aqueous or dissolved in water.

The final equation is therefore: $2Mg(s) + O_2(g) → 2MgO(s)$.

>> practice questions

1 **What is meant by sustainable development?**

2 **Why are reactions with high atom economies important?**

3 **Balance this equation and add the correct state symbols:**

$Na + H_2O → NaOH + H_2$

Chemical calculations – *Higher Tier*

 Balanced equations are used in chemical calculations.

A Reacting masses

1 **key fact** From the balanced equation for a reaction, and the mass of any one of the reactants or products, you can work out the mass of any other reactant or product.

2 For example, copper(II) oxide reacts with carbon when heated to form copper oxide and carbon dioxide. What mass of copper could be obtained from 8 g of copper(II) oxide?

Step 1	Write the equation and underline the substances in the question	$\underline{2CuO} + C$	→	$\underline{2Cu} + CO_2$
Step 2	Write the total M_r values below	2×79.5	→	2×63.5
Step 3	Show the numbers in grams	$159\,g$	→	$127\,g$
Step 4	Divide by 159 to get 1 g of CuO	$1\,g$	→	$\frac{127}{159}g$
Step 5	Multiply by 8 to get 8 g of CuO	$8\,g$	→	$\frac{127}{159} \times 8\,g = 6.4\,g$

So 6.4 g of copper could be obtained from 8 g of copper(II) oxide.

B Percentage yield

1 **key fact** The mass of a product made in a chemical reaction is called its yield. The percentage yield can be calculated:

$$\% \text{ yield} = \frac{\textbf{actual mass of product}}{\textbf{expected mass of product}} \times 100$$

2 For example, if 8 g of copper(II) oxide is reacted with excess carbon, the expected mass of copper obtained is 6.4 g.

What is the percentage yield if 5.3 g of copper is actually obtained?

$$\% \text{ yield} = \frac{\text{actual mass of product}}{\text{expected mass of product}} \times 100 = \frac{5.3}{6.4} \times 100 = 82.8\%$$

C Atom economy

1 key fact The atom economy of a chemical reaction can be calculated:

$$\% \text{ atom economy} = \frac{\text{mass of useful product}}{\text{total mass of reactants or products}} \times 100$$

2 For example, what is the atom economy of producing copper by reacting copper(II) oxide with carbon?

Step 1	Write the equation	$2CuO + C$	$\rightarrow \quad 2Cu \quad + \quad CO_2$
Step 2	Write the total M_r values below	$\rightarrow$	$2 \times 63.5 \quad 44$
Step 3	Work out toal M_r value	$\rightarrow$	$127 + 44 = 171$
Step 4	Calculate the atom economy:		

$$\% \text{ atom economy} = \frac{127}{171} \times 100 = 74.3\%$$

So 74.3% of the atoms end up in a useful product and just 25.7% are wasted as carbon dioxide.

D Empirical formulae

1 key fact The empirical formula of a compound is the simplest possible formula containing whole numbers. It can be calculated from experimental data.

2 For example, 0.92 g of sodium reacts with oxygen when heated to form 1.24 g of sodium oxide. If A_r of Na = 23 and A_r of O = 16, calculate the empirical formula of sodium oxide.

Step 1	Write down the element symbols	Na	O
Step 2	Calculate mass of each element	0.92 g	1.24 − 0.92 = 0.32 g
Step 3	Write down the A_r values below	23	16
Step 4	Divide each mass by its A_r	= 0.04	= 0.02
Step 5	Divide both by smallest number	= 2	= 1
Step 6	Write out the empirical formula	Na_2O	

>> practice questions

Use these A_r values to answer the questions: O = 16; Mg = 24; S = 32; Cu = 63.5

1 Magnesium reacts with copper oxide to form magnesium oxide:
   Mg + CuO → MgO + Cu.

   a) How much magnesium oxide could be made from 1.2 g of magnesium?

   b) What is the atom economy of the process?

2 3.2 g of sulfur reacts with 3.2 g of oxygen to produce a sulfur oxide.
   Calculate the empirical formula of the oxide.

Rates of reaction

> The rate of a reaction increases if a catalyst is used, or if one or more of the following is increased: the surface area of a solid reactant, the concentration of a dissolved reactant, the pressure of a reactant gas, or the temperature.

A Particles and collisions

1 **key fact** A chemical reaction will only happen if the reacting particles collide with each other, and have enough energy. There will be no reaction if they do not collide, or they collide with insufficient energy.

2 **key fact** The minimum amount of energy needed for a collision to be successful in causing a reaction is called the activation energy.

3 **key fact** The rate of a reaction will increase if there are more successful collisions.

B Reaction rate and surface area

1 **key fact** The rate of a chemical reaction increases if solid reactants are cut up into small pieces or made into a powder.

2 Smaller pieces have a larger surface area than larger pieces. More solid reactant particles are exposed, so there is more chance of a collision. If the rate of collisions increases, so will the rate of reaction.

C Effect of concentration and pressure

1 **key fact** The rate of a chemical reaction increases if the concentration of dissolved reactants increases, or if the pressure of a reactant gas increases.

2 The number of reactant particles in a given volume increases when:

- the concentration of a dissolved reactant increases, or
- the pressure of a reactant gas increases.

In both cases, this means that there is more chance of a collision. If the rate of collisions increases, so will the rate of reaction.

D Reaction rate and temperature

1 **The rate of a chemical reaction increases if the temperature increases.**

2 Two factors change at once when the temperature of a reactant is increased:

- the particles move more quickly, and

- a greater proportion of particles have the activation energy or more.

This means that there is more chance of a collision, and more of these collisions will be successful. If the rate of successful collisions increases, so will the rate of reaction.

E Reaction rate and catalysts

1 **key fact** **A catalyst is a substance that increases the rate of a chemical reaction without being used up in the reaction.**

2 Catalysts are important for industrial reactions. By increasing the rate of the chemical reaction, they reduce the cost of the process.

3 Different substances act as catalysts in different reactions. For example, iron is the catalyst used in the manufacture of ammonia, and nickel is the catalyst used in the hydrogenation of vegetable oils for margarine.

F Measuring rates

1 **The rate of a chemical reaction can be measured by:**
- **measuring the rate of disappearance of a reactant, or**
- **measuring the rate of appearance of a product.**

2 **key fact** **The rate of a reaction can be calculated:**

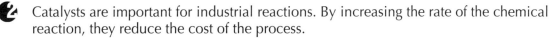

$$\text{rate of reaction} = \frac{\text{amount of reactant used or amount of product formed}}{\text{time taken}}$$

3 In a graph of amount of product formed against time, the steeper the gradient the faster the rate of reaction. If the line is horizontal, there is no reaction.

>> practice questions

1 Use the collision theory to explain what is needed for a reaction to happen.

2 Explain the effect on the rate of reaction of:

a) temperature, b) surface area, c) concentration.

Reversible reactions

A Reversible reactions

1 key fact A reversible reaction is a reaction in which the products can react together to form the reactants again.

2 key fact In a chemical equation to show a reversible reaction, the symbol $\rightleftharpoons$ is used instead of an arrow. For example:

ammonium chloride $\rightleftharpoons$ ammonia + hydrogen chloride

This shows that ammonium chloride can decompose to form ammonia and hydrogen chloride, and that these two substances can react together to make ammonium chloride again. Both reactions happen at the same time.

B Energy changes in reactions

1 key fact Energy is transferred to or from the surroundings when chemical reactions happen. This is usually as heat, giving a temperature change:

- exothermic reactions transfer heat to the surroundings, giving an increase in temperature

- endothermic reactions take in heat from the surroundings, giving a decrease in temperature.

2 Neutralisation and combustion are examples of exothermic reactions. Thermal decomposition and electrolysis are examples of endothermic reactions.

3 key fact Reversible reactions are exothermic in one direction and endothermic in the other direction. The same amount of energy is involved. The only difference is whether it is given out to the surroundings or taken in from them.

4 White anhydrous copper(II) sulfate can be used as a test for water. It turns blue when it reacts with water, but this is a reversible reaction:

$$\text{anhydrous copper(II) sulfate + water} \underset{\text{endothermic}}{\overset{\text{exothermic}}{\rightleftharpoons}} \text{hydrated copper(II) sulfate}$$
$$\text{(white)} \qquad\qquad\qquad\qquad\qquad\qquad\qquad\qquad \text{(blue)}$$

If the blue hydrated copper(II) sulfate is heated, the water is driven off to leave white anhydrous copper(II) sulfate again.

C Equilibria – *Higher Tier*

1 **key fact** If a reversible reaction happens in a closed system, such as a container with the lid on, it reaches equilibrium.

2 **key fact** At equilibrium, the forward and reverse reactions are still happening. But they happen at the same rate, so the concentrations of all the substances involved stay the same.

3 **key fact** The position of equilibrium describes the relative amounts of the substances on each side of the equation.
For example, look at the reversible reaction:

ammonium chloride ⇌ ammonia + hydrogen chloride

The position of equilibrium lies to the left if there is more ammonium chloride than ammonia and hydrogen chloride.

4 The position of equilibrium can be changed by altering the reaction conditions. This is important for finding the best conditions for industrial reactions involving reversible reactions.

>> practice questions

1 When water is added to anhydrous copper(II) sulfate, will heat energy be given off or taken in? Explain your answer.

2 In the reversible reaction $N_2(g) + 3H_2(g) \rightleftharpoons 2NH_3(g)$, the forward reaction is exothermic. What can be said about the reverse reaction?

The Haber process

Ammonia is made by reacting nitrogen and hydrogen at 450 °C and 200 atmospheres of pressure, in the presence of an iron catalyst.

These conditions are chosen to achieve a balance between the rate of reaction and yield of ammonia.

A Ammonia

1 Ammonia NH_3 is a colourless gas with a sharp, characteristic smell. It is very soluble in water, and it acts as a weak base.

2 Ammonia and its products are widely used as artificial fertilisers. Ammonium nitrate NH_4NO_3 is a good source of nitrogen for crops.

B The Haber process

1 **key fact** Ammonia is made by reacting nitrogen with hydrogen using the Haber process. The reaction is reversible:

nitrogen + hydrogen $\rightleftharpoons$ ammonia

$N_2(g) + 3H_2(g) \rightleftharpoons 2NH_3(g)$

2 **key fact** The raw materials for the Haber process are air, water, and natural gas. Coal is sometimes used instead of natural gas.

3 Hydrogen for the Haber process is made by reacting natural gas with steam:
$CH_4(g) + 2H_2O(g) \rightleftharpoons CO_2(g) + 4H_2(g)$

4 Nitrogen for the Haber process is made by the fractional distillation of air. It is also made by burning hydrogen in air. This removes the oxygen, leaving almost pure nitrogen behind.

5 **key fact** Nitrogen and hydrogen are passed over an iron catalyst at about 450 °C and 200 atmospheres of pressure.

6 **key fact** The reaction mixture is cooled. This makes the ammonia liquefy and it is removed. Unreacted nitrogen and hydrogen are recycled.

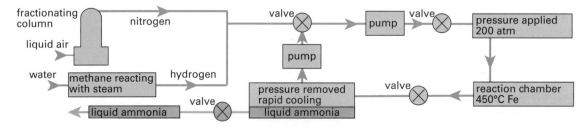

C Pressure and the Haber process – *Higher Tier*

1 **key fact** In reversible reactions involving reacting gases, the position of equilibrium moves in the direction of the fewest molecules of gas if the pressure is increased.

2 **key fact** In the Haber process, $N_2(g) + 3H_2(g) \rightleftharpoons 2NH_3(g)$, there are $1 + 3 = 4$ molecules of gas on the left of the equation and only 2 on the right. As the pressure is increased, the position of equilibrium moves further to the right.

3 The higher the pressure, the greater the yield of ammonia. High pressures are expensive to maintain, so the actual pressure chosen is a compromise.

D Temperature and the Haber process – *Higher Tier*

1 **key fact** In reversible reactions, the position of equilibrium moves in the direction of:
- the endothermic reaction if the temperature is increased
- the exothermic reaction if the temperature is decreased.

2 **key fact** In the Haber process, $N_2(g) + 3H_2(g) \rightleftharpoons 2NH_3(g)$, the forward reaction is exothermic. As the temperature is decreased, the position of equilibrium moves further to the right.

3 The lower the temperature, the greater the yield of ammonia. But at low temperatures the rate of reaction is low. So the actual temperature chosen is a compromise. It is high enough to get a reasonable rate of reaction, but low enough to achieve a reasonable yield of ammonia.

>> practice questions

1 Explain why the manufacture of artificial fertilisers uses fossil fuels.

2 Give three ways in which the reactions in the Haber process are made to run quickly.

Electrolysis

- Electrolysis is the breaking down of ionic substances using an electric current.

- It is used industrially to produce useful substances from sodium chloride solution and to purify copper.

A Electrodes and electrolysis

① An electrode is usually made from graphite or an unreactive metal, such as copper or platinum. Two electrodes are needed for electrolysis: a negative electrode or cathode, and a positive electrode or anode.

② **key fact** Ionic substances conduct electricity when they are dissolved in water or melted. This is because their ions are free to move in the solution or liquid.

③ **key fact** When electricity is passed through an ionic substance that is in solution or molten, the substance is decomposed into elements. This is electrolysis.

B At the electrodes

① **key fact** Positive ions move to the negative electrode:

- they gain electrons and are reduced
- for example, $Cu^{2+}(aq) + 2e^- \rightarrow Cu(s)$.

② **key fact** Negative ions move to the positive electrode:

- they lose electrons and are oxidised
- for example, $2Cl^-(aq) \rightarrow Cl_2(g) + 2e^-$.

exam tip >>

Remember '*OIL-RIG*'. Oxidation *is l*oss of electrons. Reduction *is g*ain of electrons.

C Mixtures of ions

1 key fact When a mixture of ions is electrolysed, the products that form depend upon the reactivity of the elements involved.

2 In a solution containing metal ions, hydrogen ions from the water will be present, too.
At the negative electrode:

- hydrogen gas is given off if the metal is more reactive than hydrogen

- the metal is given off if it is less reactive than hydrogen.

3 In a solution, oxygen is given off at the positive electrode unless Cl⁻, Br⁻ or I⁻ ions are present.
In these cases Cl_2, Br_2 or I_2 are given off instead.

most reactive potassium
sodium
calcium
magnesium
aluminium
zinc
iron
tin
lead
hydrogen
copper
silver
gold
least reactive platinum

The reactivity series shows you the relative reactivity of hydrogen and metals. These produce positive ions.

D Industrial use of electrolysis

1 key fact The electrolysis of sodium chloride solution produces hydrogen at the negative electrode, and chlorine at the positive electrode. Sodium hydroxide solution is also produced.

2 Hydrogen is used as a fuel and in the manufacture of margarine. Chlorine is used to manufacture bleach and some plastics.
Sodium hydroxide is used in the manufacture of soap and paper.

3 key fact Copper is purified by electrolysis. The positive electrode is impure copper, and the negative electrode is pure copper. Electricity is passed through a solution of Cu^{2+} ions, such as copper(II) sulfate.

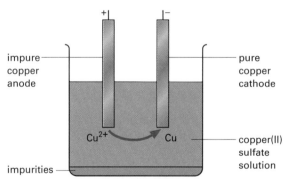

impure copper anode — pure copper cathode — copper(II) sulfate solution — impurities — Cu^{2+} — Cu

>> practice questions

1 **During the electrolysis of dilute sulfuric acid, hydrogen and oxygen are produced.**

  a) **Which electrode is each gas released from?**

  b) **Are hydrogen ions oxidised or reduced when they reach the electrode?**

2 **What substance is released at the negative electrode during the electrolysis of:**

  a) **potassium chloride solution,**

  b) **copper chloride solution?**

Acids, bases and neutralisation

> ⟨image⟩ **Acids release hydrogen ions in solutions and alkalis release hydroxide ions.**
>
> ⟨image⟩ **In neutralisation reactions, hydrogen ions react with hydroxide ions to make water.**
>
> ⟨image⟩ **The particular salt made in a neutralisation reaction depends on the acid and alkali used.**

A The pH scale

① **key fact** The pH scale is a measure of the acidity or alkalinity of a solution. It runs from 0 to 14:

- acids have a pH of less than 7
- neutral solutions have a pH of 7
- alkalis have a pH of more than 7.

② The pH of a solution can be found using universal indicator and a colour chart, or using a pH meter.

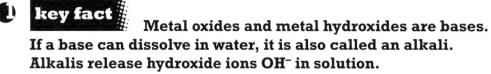

pH

0 1 2 3 4 5 6 7 8 9 10 11 12 13 14

← increasingly acidic   neutral   increasingly alkaline →

red   orange   green   blue   purple

B Acids

① **key fact** Acids release hydrogen ions H+ in solution.

② The stronger the acid, the more hydrogen ions it releases and the lower its pH.

③ Common strong acids include nitric acid HNO_3, hydrochloric acid HCl and sulfuric acid H_2SO_4.

C Bases and alkalis

① **key fact** Metal oxides and metal hydroxides are bases. If a base can dissolve in water, it is also called an alkali. Alkalis release hydroxide ions OH^- in solution.

② The stronger the alkali, the more hydroxide ions it releases and the higher its pH.

③ Sodium hydroxide NaOH is a common strong alkali and ammonia solution is a weak alkali.

D Neutralisation

1 **key fact** In neutralisation reactions, hydrogen ions from acids react with hydroxide ions from alkalis to produce water:

$$H^+(aq) + OH^-(aq) \rightarrow H_2O(l)$$

2 The reaction between hydrogen ions and hydroxide ions is exothermic, so heat is given off when acids react with alkalis.

3 A salt is also produced in the reaction between an acid and a base.

E Naming salts

1 **key fact** The salt produced in a reaction between an acid and a base depends upon the particular acid and base used.

2 **key fact** The first part of the salt's name comes from the metal in the base or alkali. If the alkali is ammonia solution, this part is 'ammonium'.

3 **key fact** The second part of the salt's name comes from the acid.

acid	second part of salt's name
hydrochloric acid	chloride
nitric acid	nitrate
sulfuric acid	sulfate

4 For example, sodium hydroxide and hydrochloric acid react to produce sodium chloride. Ammonia solution and nitric acid react to produce ammonium nitrate.

>> practice questions

1 Why are all alkalis bases, but only some bases are alkalis?

2 a) Which ions are released by acids and bases?

   b) What happens during a neutralisation reaction?

Making salts

- Different methods are needed to make salts, depending on whether the salt is insoluble or soluble.

- Solid salts can be produced from salt solutions by crystallisation.

A Insoluble salts

1 **key fact** Insoluble salts do not dissolve in water. They can be produced by precipitation and separated from the reaction mixture by filtration.

2 Precipitation reactions happen when two solutions of ions are mixed and an insoluble salt or precipitate is produced.

3 **key fact** Precipitation reactions are useful for treating drinking water to remove unwanted ions from solution.

4 It helps to know which ions are soluble and which are not.

soluble	insoluble
most chlorides, bromides and iodides	silver chloride, bromide and iodide lead chloride, bromide and iodide
most sulfates	lead sulfate and barium sulfate calcium sulfate is slightly soluble
all sodium, potassium and ammonium salts	
all nitrates	

5 For example, silver chloride is insoluble.
You need a soluble silver salt and a soluble chloride to make it:

- all nitrates are soluble, so silver nitrate is soluble

- all sodium salts are soluble, so sodium chloride is soluble.

A precipitate of solid silver chloride is made when they are mixed:

silver nitrate + sodium chloride → silver chloride + sodium nitrate
$AgNO_3(aq) + NaCl(aq) \rightarrow AgCl(s) + NaNO_3(aq)$

exam tip >>

Be prepared to suggest how to make a named salt in the exam.

64

B Soluble salts

1 **key fact** Soluble salts can be made by reacting acids with metals, insoluble bases or alkalis. The salt solution is warmed gently to evaporate the water, leaving solid salt behind. This is called crystallisation.

2 **key fact** The method chosen depends upon the particular salt that is needed.

3 Metals react with acids to produce salts and hydrogen:

metal + acid → salt + hydrogen

This does not work if the metal is less reactive than hydrogen, and it is too dangerous if the metal is very reactive, like sodium or potassium.

4 Insoluble bases react with acids to produce salts and water:

base + acid → salt + water

This is useful if a metal cannot be used. For example, copper does not react with sulfuric acid. But copper(II) oxide does:

copper(II) oxide + sulfuric acid → copper(II) sulfate + water

The excess, unreacted copper(II) oxide can be filtered off leaving a solution of copper(II) sulfate.

5 Alkalis react with acids to produce salts and water:

alkali + acid → salt + water

This is also useful if a metal cannot be used. For example, it is too dangerous to react sodium with hydrochloric acid. Sodium hydroxide solution is used instead:

sodium hydroxide + hydrochloric acid → sodium chloride + water

The reaction is followed using an indicator such as universal indicator to see when the reaction mixture has become neutral.

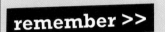

remember >>

A lighted wooden splint makes a popping sound when it is held in a test tube of hydrogen.

>> practice questions

1 Name two solutions that would react together to make insoluble lead iodide.

2 Suggest why it is better to make potassium nitrate using potassium hydroxide and nitric acid, rather than potassium and nitric acid.

Speed, distance and acceleration

> The speed of an object is the rate that its distance from a point changes.
>
> The velocity of an object is its speed in a particular direction.
>
> The acceleration of an object is its rate of change in velocity.

A Distance and speed

① key fact If an object moves in a straight line, its speed can be worked out:

$$\text{speed (m/s)} = \frac{\text{change in distance (m)}}{\text{time taken for change (s)}}$$

② key fact A distance-time graph has the distance on the vertical axis and the time on the horizontal axis. Its gradient represents the speed. The greater the gradient, the greater the speed of the object.

A horizontal line on a distance-time graph represents a stationary object.

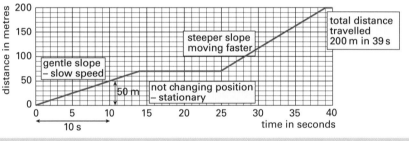

- total distance travelled 200 m in 39 s
- steeper slope moving faster
- gentle slope – slow speed
- not changing position – stationary
- 50 m
- 10 s

Higher Tier

❸ You need to be able to work out the speed of an object by working out the gradient of a distance-time graph.
For example, between 0 s and 10 s the object moved from 0 m to 50 m.

$$\text{speed (m/s)} = \frac{\text{change in distance (m)}}{\text{time taken for change (s)}} = \frac{(50-0)}{(10-0)} = \frac{50}{10} = 5\,\text{m/s}$$

B Velocity

① key fact The velocity of an object is its speed in a given direction.

② key fact The velocity of an object changes when it:

- travels at the same speed but changes direction
- travels in the same direction but changes speed
- changes speed and direction.

An object in orbit changes velocity, even if its speed stays the same.

C Acceleration

① key fact An object accelerates if it changes velocity:

$$\text{acceleration (m/s}^2) = \frac{\text{change in velocity (m/s)}}{\text{time taken for change (s)}}$$

② key fact A velocity-time graph has the velocity on the vertical axis and the time on the horizontal axis. Its gradient represents the acceleration. The greater the gradient, the greater the acceleration of the object.

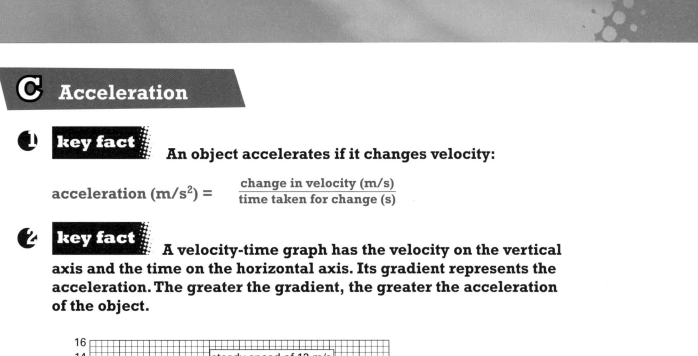

A horizontal line on a velocity-time graph represents an object moving at constant velocity.

Higher Tier

③ You need to be able to work out the acceleration of an object.
You can do this by working out the gradient of a velocity-time graph.
For example, between 0 s and 20 s the velocity of the object changed from 0 m/s to 20 m/s.

$$\text{acceleration (m/s}^2) = \frac{\text{change in velocity (m/s)}}{\text{time taken for change (s)}} = \frac{(12 - 0)}{(20 - 0)} = \frac{12}{20}$$

$$= 0.6\,\text{m/s}^2$$

④ You need to be able to work out the distance travelled by an object. You can do this by working out the area under the graph.
For example, what is the distance travelled between 40 s and 50 s?

$$\text{area of a triangle} = \frac{1}{2} \times \text{base} \times \text{height} = \frac{1}{2} \times (50 - 40) \times 12 = 60\,\text{m}$$

>> practice questions

1 A person travels 100 m in 20 s. What is their mean speed?

2 A car increases in speed from 15 m/s to 30 m/s in 30 s. What is its acceleration?

3 A train travels at 50 m/s for 20 s. How far does it travel?

Forces

- If the forces on an object are balanced, the object remains at rest or moves at a constant speed in a straight line.

- An unbalanced force on an object causes it to accelerate or change direction.

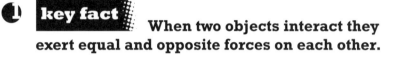

A Resultant force

1 key fact When two objects interact they exert equal and opposite forces on each other.

When a sprinter pushes on the starting block at the start of a race, the block pushes back with the same force.

 2 key fact All the forces acting on an object can be described by a single force that has the same effect as all the forces acting together. This is called the resultant force.

3 The effect of the resultant force on an object depends upon:

- whether the resultant force is zero or not, and
- whether the object is moving or stationary.

B Balanced forces

 1 key fact When all the forces acting on an object cancel each other out, the resultant force is zero. The forces are balanced forces.

2 If an object is stationary and the resultant force on it is zero, the object stays still.

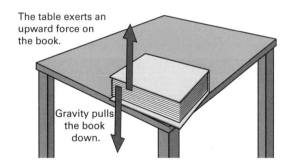

The table exerts an upward force on the book.

Gravity pulls the book down.

The weight of the stationary book is balanced by the upward force from the table.

If an object is moving and the resultant force on it is zero, the object keeps moving at the same speed and in the same direction.

The force from the car's engine is balanced by air resistance and other frictional forces.

C Unbalanced forces

key fact When all the forces acting on an object do not cancel each other out, the resultant force is not zero. The forces are unbalanced forces.

If an object is stationary and the resultant force on it is not zero, the object will accelerate in the direction of the resultant force.

Unbalanced forces make this shopping trolley begin to move backwards.

If an object is moving and the resultant force on it is not zero, the object accelerates in the direction of the resultant force.

The force of the engine in this racing car is not balanced by air resistance. The car accelerates forwards.

>> practice questions

1 **Look at the diagram of the lorry.**

a) **What can you say about the resultant force?**

b) **Explain what will happen to the lorry.**

Weight and falling

> - **Objects fall because of gravity.**
> - **When objects fall through air, there is also air resistance.**
> - **Falling objects reach a steady speed when their weight is balanced by the air resistance.**

A Force, mass and acceleration

1 **key fact** The unit of force is the newton, N.

2 **key fact** Force, mass and acceleration are related by this equation:

resultant force (N) = mass (kg) × acceleration (m/s^2)

3 When an unbalanced force makes an object accelerate:

- the greater the force, the greater the acceleration, and
- the greater the mass, the smaller the acceleration.

It is easier to push a wheelbarrow full of leaves than one full of heavy stones.

4 For example, the braking force on a car gives a deceleration of 2 m/s^2.

The mass of the car is 1500 kg. Calculate the braking force.

resultant force $\quad$ = mass × acceleration
$\qquad\qquad\qquad\quad$ = 1500 × 2 N
$\qquad\qquad\qquad\quad$ = 3000 N

B Weight

1 Gravity is a force that attracts all objects towards each other. Very large objects like the Earth have very powerful forces of gravity. The Earth's gravity attracts objects towards its centre. The force on an object due to gravity is its weight.

2 The weight of an object depends on the mass of the object and the gravitational field strength.

3 **key fact** The unit of weight is the newton, N.

④ **key fact** The weight of an object can be calculated using this equation:

weight (N) = mass (kg) × gravitational field strength (N/kg)

⑤ **key fact** The Earth's gravitational field strength is approximately 10 N/kg.

⑥ **key fact** For example, a bar of chocolate has a mass of 0.1 kg. What is its weight on Earth?

weight = mass × gravitational field strength
        = 0.1 × 10 N
        = 1.0 N

C Falling

① Liquids and gases are fluids. When objects fall through a fluid, frictional forces act in the opposite direction to the movement.

② In air, these frictional forces are called air resistance.

③ The size of air resistance depends upon the shape and speed of the object:

- the greater the surface area, the greater the air resistance
- the faster the object moves, the greater the air resistance.

④ **key fact** An object falling through air experiences two forces, its weight acting downwards, and air resistance acting upwards.

⑤ **key fact** As the object falls it accelerates and the air resistance increases. Eventually the weight is balanced by the air resistance. The resultant force becomes zero and the object falls at a steady speed, called its terminal velocity.

A falling object accelerates until it reaches its terminal velocity.

>> practice questions

1 **Look at the diagrams of the parachutist.**

a) **How do the parachutist's weight and air resistance change as she falls?**

b) **What happens to her speed as she falls?**

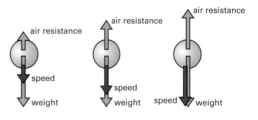

Stopping distances

The distance a car travels while the brakes are applied depends on the mass of the car, its speed and the braking force. It can be increased by poor weather and worn brakes.

The distance a car travels before it stops depends on the speed of the car, the speed of the driver's reactions and the braking distance.

A Background

1. A car will travel at a steady speed if the frictional forces balance the driving force from the engine. Once the brakes are applied, they cause additional frictional forces that eventually stop the car.

2. **key fact** The distance needed to stop the car from seeing a hazard to the car being stationary is its stopping distance.

3. **key fact** The stopping distance depends on the distance travelled by the car:
   - **during the driver's reaction time, and**
   - **once the brakes have been applied.**

4. **key fact** stopping distance = thinking distance + braking distance

B Thinking distance

1. When a driver sees a hazard, it takes a certain amount of time to react to it and apply the brakes. The time taken is the reaction time.

2. During the reaction time, the car continues to travel. The distance it goes during this time is the thinking distance.

3. **key fact** A driver's reaction time is increased if they:
   - **are tired**
   - **under the influence of drugs or alcohol.**

   These factors increase the thinking distance.

4. A driver's reaction time may also be increased if the weather conditions are poor, making it difficult to see ahead. Distractions such as using a mobile phone, or carrying badly behaved children in the car, can also increase a driver's reaction time.

C Braking distance

1 **key fact** The braking distance is the distance travelled by the car once the brakes have been applied.

2 **key fact** It depends upon factors such as the mass of the car, its speed, its condition, and the road and weather conditions.

3 The more mass a car has, the smaller the deceleration for a given braking force and the greater its braking distance. A fully loaded car will travel further under braking than one just containing its driver.

4 The faster a car is travelling, the further it will travel under braking before it stops. This is why it is important to keep within the speed limit for the road.

5 A poorly maintained car is likely to have a greater braking distance than a well maintained one under the same conditions. Worn tyres have less friction on the road surface and the car may skid. Worn brakes produce a weaker braking force, so the car will travel further under braking before it stops.

6 There is less friction between a car's tyres and the road surface when the road is wet or icy than when it is dry. The braking distance increases under such conditions, and the car may skid if the brakes are applied too hard.

exam tip >>

Be prepared to explain the factors involved in the stopping distance, and to use information given to you to calculate it.

>> practice questions

1 Explain the effect on the stopping distance of a car if:

a) the car is travelling very quickly,

b) the car's brakes are worn,

c) the road is icy.

2 a) A car is travelling at 15 m/s. The driver's reaction time is 0.6 s. What is the thinking distance?

b) If the braking distance is 16.7 m, what is the overall stopping distance of the car?

Work and kinetic energy

- Work is done when a force moves an object.
- The kinetic energy of an object depends upon its mass and speed.

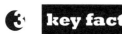 A Work

1 key fact When a force is applied to an object and makes it move, energy is transferred and work is done. For example, you do work if you push your bike along the pavement, or lift a book onto a bookshelf.

2 key fact Energy and work are both measured in joules, J.

3 key fact You can use this equation to calculate the work done by a force:

work done (J) =
force applied (N) × distance moved in direction of force (m)

4 For example, a weightlifter lifts a 200 kg mass by 1.5 metres. How much work does he do?

200 kg has a weight of 200 × 10 = 2000 N
work done = force × distance = 2 000 × 1.5 = 3000 J

remember >>

weight = mass × gravitational field strength. The gravitational field strength on Earth is approximately 10 N/kg. So on Earth, weight (N) = mass (kg) × 10.

B Kinetic energy

1 key fact Kinetic energy is the scientific name for movement energy. The amount of kinetic energy an object has depends upon its mass and speed:

- If two objects travel at the same speed, the object with the greatest mass has the most kinetic energy.

- If two objects have the same mass, the object with the greatest speed has the most kinetic energy.

A car has more kinetic energy than a bike travelling at the same speed.

Kinetic energy can be transformed into other types of energy. For example:

- When the brakes on a moving car are applied, kinetic energy is transformed into heat energy. If the brakes are squeaky it is transformed into sound energy, too.

remember >>

The work done against frictional forces, such as those in car brakes, is mainly transformed as heat energy.

Higher Tier

 key fact You can use this equation to calculate the kinetic energy of an object:

kinetic energy (J) = $\frac{1}{2}$ × mass (kg) × speed^2 (m/s^2)

④ For example, an 80 kg man runs at 5 m/s. What is his kinetic energy?

kinetic energy = $\frac{1}{2}$ × mass × speed^2

= $\frac{1}{2}$ × 80 × 5^2

= $\frac{1}{2}$ × 80 × 25 = 1000 J

C Elastic potential energy

 key fact **When an object is squashed or stretched, elastic potential energy is stored in the object when work is done to change its shape. This happens if the object can go back to its original shape.**

② For example, a squash ball stores and releases elastic potential energy every time it hits the wall. The strings in the squash racquet store and release elastic potential energy every time the ball hits them.

>> practice questions

1 How much work is done when a 1.5 N chocolate bar is moved 2 m along a table?

2 Which is likely to have the most kinetic energy at 5 m/s, a golf ball or a table tennis ball?

3 Explain why a beach ball can store elastic potential energy but a stone cannot.

4 Calculate the kinetic energy of a 10 kg mass moving at 4 m/s.

75

Momentum

- The momentum of an object depends upon its mass and velocity.

- Momentum is conserved in a collision or explosion.

- Cars have safety features to reduce the forces on the passengers in an accident.

A The basics

1 Momentum is a measure of the tendency of a moving object to keep moving in the same direction.

2 **key fact** A force must be applied to an object that can move, or to a moving object, to change its momentum.

3 **key fact** You can calculate the momentum of an object using this equation:

momentum (kg m/s) = mass (kg) × velocity (m/s)

4 For example, what is the momentum of a 10 kg mass moving at 4 m/s?
momentum = 10 × 4 = 40 kg m/s

B Conservation of momentum

1 Momentum has direction because velocity, not speed, is involved. Two identical objects travelling in opposite directions will have the same size momentum, but one will have a positive value and the other will have a negative value.

2 **key fact** Momentum is conserved in any collision or explosion if no external force acts on the colliding or exploding objects:

- **The total momentum is the same before and after the collision or explosion.**

3 For example, a 1000 kg car travelling at 5 m/s bumps into another 1000 kg car at traffic lights. After the collision, their bumpers get stuck and they move off together. What velocity do the cars travel at?

15 m/s, 1000 kg

0 m/s, 1000 kg

before collision

total momentum = momentum of moving car + momentum of stationary car

$$= (1000 \times 5) + (1000 \times 0) \, \text{kg m/s}$$

$$= 5000 + 0 = 5000 \, \text{kg m/s}$$

after collision

total momentum = 5000 kg m/s

total mass = 2000 kg

new velocity = momentum ÷ mass

$$= 5000 \div 2000 = 2.5 \, \text{m/s}$$

C Change in momentum – *Higher Tier*

1 **key fact** The force needed to change the momentum of an object can be calculated using this equation:

$$\text{force (N)} = \frac{\text{change in momentum (kg m/s)}}{\text{time taken for change (s)}}$$

2 For example, what force is needed to change the speed of a 1000 kg car by 5 m/s in 4 s?

change in momentum = $1000 \times (5 - 0) = 5000 \, \text{kg m/s}$

$$\text{force (N)} = \frac{\text{change in momentum (kg m/s)}}{\text{time taken for change (s)}} = \frac{5000}{4} = 1250 \, \text{N}$$

D Car safety features

1 Cars have a number of features that reduce the forces experienced by passengers in a crash. They work by increasing the time taken for their momentum to become zero.

2 Safety features in cars include crumple zones, air bags and seat belts.

>> practice questions

1 What is the momentum of a 0.06 kg tennis ball travelling at 50 m/s?

2 A 50 kg person steps off a 100 kg boat. The person moves off the boat at 1 m/s. How fast does the boat move in the opposite direction?

3 What force is needed to change the momentum of a boat by 25 kg m/s in 5 s?

Static electricity

A Electrostatic charge

1 Electrons are negatively charged.

2 **key fact** When some materials are rubbed together they become electrically charged:

- one material gains electrons and becomes negatively charged
- the other material loses electrons and becomes positively charged.

If the materials are insulators, the charge does not leak away. The materials have electrostatic charge.

3 **key fact** Electrically charged objects exert a force on each other:

- objects with like charges repel each other
- objects with opposite charges attract each other.

When you brush your hair, the hairs and brush become oppositely charged and attract each other. The hairs repel each other because they have like charges.

B Using electrostatic charge

1 **key fact** Photocopiers and laser printers use static electricity. Negatively charged particles of black toner are attracted to positively charged areas of a drum in the photocopier. The toner is transferred to the sheet of paper, which is pressed and heated to make the toner stick tightly to it.

key fact Electrostatic precipitators remove ash from power station waste gases. Metal electrodes inside the chimney attract ash particles, which are charged.

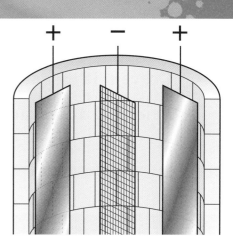

C Dangers of static electricity

① A charged object can discharge with a spark. This is why you may see tiny flashes of light in the dark when you take your pullover off.

② **key fact** A charged object can be discharged by connecting it to earth with a conductor, such as a metal cable.

③ Static discharge can be dangerous. Fuel passing along a refuelling pipe causes enough friction to give the pipe an electrostatic charge. If the charge builds up there is the danger of a spark that may cause an explosion.

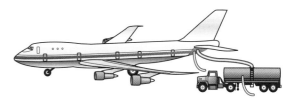

Higher Tier

④ The greater the charge on an object, the greater the potential difference (voltage) there is between the object and earth. If the potential difference is high enough, a spark can jump from the object to an earthed conductor. This is why you may get shocks off handrails as you walk upstairs.

exam tip >>

Be prepared to explain why static electricity can be dangerous, and how it can be discharged safely.

>> practice questions

1 Explain why a balloon sticks to your hair after you rub it on a pullover.

2 When a tanker delivers fuel, a wire links it to earth. Explain how this helps to prevent an explosion.

Electrical circuits

A Representing circuits

1. Circuit diagrams are used to show electrical circuits. Standard circuit symbols are used in these diagrams.

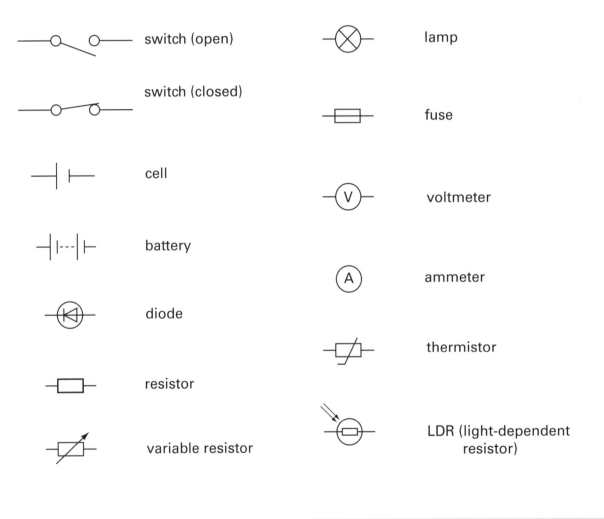

switch (open)

switch (closed)

cell

battery

diode

resistor

variable resistor

lamp

fuse

voltmeter

ammeter

thermistor

LDR (light-dependent resistor)

exam tip >>

You must be able to draw, recognise and use these circuit symbols in circuit diagrams.

B Current

1 **key fact** The rate of flow of electrical charge is called the current.

2 **key fact** Current is measured in amperes, A, often just called amps.

3 **key fact** The greater the current flowing through a component, the greater the rate of flow of electrical charge.

4 **key fact** The current through a component is measured using an ammeter in series with the component.

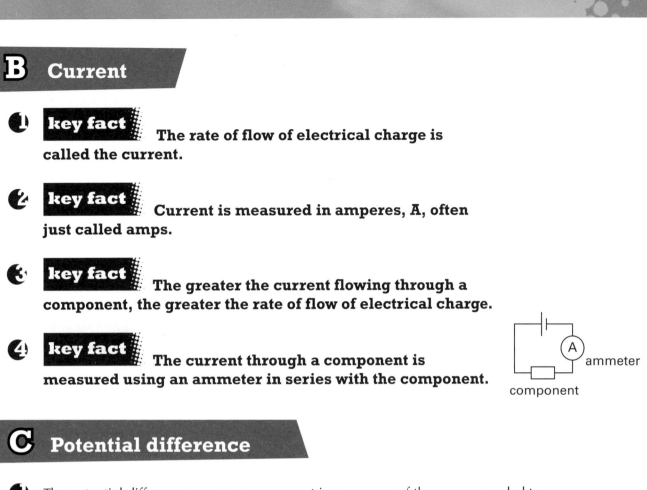

C Potential difference

1 The potential difference across a component is a measure of the energy needed to move electrical charge through it.

2 **key fact** Potential difference is measured in volts, V. It is often called the voltage.

3 **key fact** The potential difference across a component is measured using a voltmeter in parallel with the component.

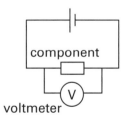

>> practice questions

1 Draw a circuit diagram for a cell, lamp and open switch in series.

2 Explain how current and potential difference are measured.

Resistance

 Resistance can be calculated if you know the potential difference and current.

 The resistance of a component varies depending on the type of component it is.

A Resistance

1. The resistance in a component makes it difficult for electric current to flow through. The greater the resistance, the more difficult it is for current to flow.

2. Resistance is measured in ohms, Ω.

3. Wires usually have low resistance:

   • the shorter the wire, the lower the resistance

   • the thicker the wire, the lower the resistance.

4. Some components have high resistance. These are called resistors. The resistance of variable resistors can be altered to control the current in a circuit.

remember >>

Variable resistors are often used in control knobs and sliders for volume and brightness.

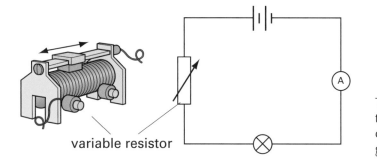

variable resistor

The longer the wire, the greater the resistance and the smaller the current. A lamp in the circuit will glow less brightly.

B Calculating resistance

1. **key fact** **Potential difference, current and resistance are related:**

   potential difference (V) = current (A) × resistance (Ω)

   The greater the resistance, the smaller the current through a component for a given potential difference.

2. You can calculate the resistance in a component by rearranging the equation:

   $$\text{resistance (}\Omega\text{)} = \frac{\text{potential difference (V)}}{\text{current (A)}}$$

3. For example, what is the resistance of a lamp filament if the potential difference is 12V and the current is 3 A?

   $$\text{resistance} = \frac{\text{potential difference}}{\text{current}} = \frac{12}{3} = 4\,\Omega$$

C Current-potential difference graphs

1 **key fact** Current-potential difference graphs show how the current flowing through a component changes as the potential difference across it changes. Different components produce different shaped lines.

2 The current through a resistor is directly proportional to the potential difference, as long as the temperature stays the same.

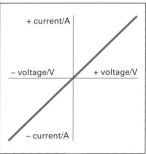

Wires and resistors give this sort of graph.

3 The resistance through a filament lamp increases as the temperature increases. This makes the current-potential difference graph level off at high potential differences.

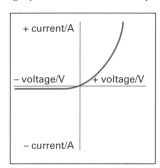

The graph for a filament lamp.

4 Diodes have a very high resistance in one direction. This means that current only flows through a diode in the opposite direction, where the resistance is much less.

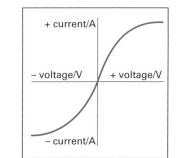

The graph for a diode.

5 The resistance of an LDR (light-dependent resistor) decreases as the light intensity decreases. The resistance of a thermistor decreases as the temperature increases.

>> practice questions

1 What is the resistance of a 5A lamp at 250V?

2 Describe the current-potential difference graphs for:

   a) a resistor at constant temperature,

   b) a filament lamp,

   c) a diode.

Parallel and series circuits

 Parallel and series circuits have different characteristics in terms of the potential difference across their components, and the current flowing through their components.

A Cells in series

① key fact When cells are connected in series in the same direction, the total potential difference they provide is the sum of the individual potential differences. The potential difference of any cells connected in the opposite direction is subtracted from the total.

② For example, four 1.5V cells are connected in the same direction.
The total potential difference they provide is $(4 \times 1.5) = 6.0$V.
If one of the cells is connected in the opposite direction, the total potential difference they provide is $(3 \times 1.5) - 1.5 = 3.0$V.

B Potential difference

① key fact When components in a circuit are connected in parallel, the potential difference across each of them is the same.

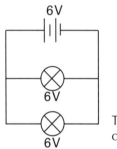

The potential difference across all the components in this parallel circuit is 6 V.

② When components in a circuit are connected in series, the potential difference of the supply is shared between them.

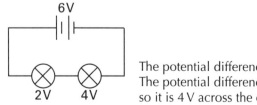

The potential difference of the supply is 6V.
The potential difference across one lamp is 2 V, so it is 4 V across the other.

C Current

① key fact When components in a circuit are connected in parallel, the total current flowing through the whole circuit is the sum of all the currents through the individual components.

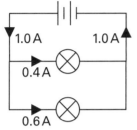

In a parallel circuit, the current flowing into a junction is the same as the current flowing out of it.

② When components in a circuit are connected in series, the same current flows through all the components.

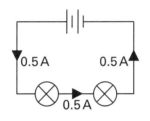

In a series circuit, the current is the same anywhere in the circuit.

remember >>

When components are connected in series, the total resistance is equal to the individual resistances added together.

>> practice questions

1 Study the circuit diagram below.

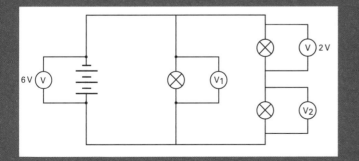

a) If each cell is identical, what is the potential difference provided by one of them?

b) What is the voltage V1?

c) What is the voltage V2?

Mains electricity

- **Electricity can be supplied as d.c. or a.c.**
- **Mains electricity is an a.c. supply.**
- **Cables and plugs have features to make them safe to use.**

A a.c. and d.c.

① key fact Direct current or d.c. always passes in the same direction. The output from cells and batteries is d.c.

potential difference, V

—————————————— d.c.
—————————————— time/s

The oscilloscope trace for a d.c. current.

② key fact Alternating current or a.c. changes direction constantly. The output from mains electricity supply is a.c.

potential difference, V

a.c.

time/s

The oscilloscope trace for an a.c. current.

B The UK mains supply

>> key fact The UK mains electricity is:

- an a.c. supply with a frequency of 50 Hz (50 cycles per second)
- supplied to households at about 230V.

C The UK mains supply – *Higher Tier*

① key fact You can work out the frequency of the mains supply by looking at its oscilloscope trace. There is one complete cycle every 1/50th of a second, so the frequency is 50 cycles per second or 50 Hz.

potential difference, V

a.c.

time/s

$\frac{1}{50}$ $\frac{2}{50}$ $\frac{3}{50}$

2 key fact When the live terminal is positive, the neutral terminal is negative. And when the live terminal is negative, the neutral terminal is positive.

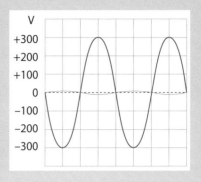

The red trace is from the live terminal and the blue trace is from the neutral terminal. Notice the peak voltage of the live terminal is actually more than 230V.

D The three-pin plug

1 key fact Electrical devices are usually connected to the mains supply using a cable and a three-pin plug.

2 key fact The cable contains two or three inner wires. Each wire has a copper core because copper is a good conductor of electricity. The outer layers are plastic because this is a good electrical insulator. The inner wires are colour coded.

3 The three-pin plug has several key features:

- a case made from tough plastic or rubber
- three pins made from brass
- a fuse between the live terminal and the live pin
- a cable grip to secure the cable.

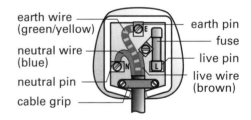

earth wire (green/yellow) — earth pin — fuse
neutral wire (blue) — live pin
neutral pin — live wire (brown)
cable grip

E Fuses and circuit breakers

1 A circuit should be switched off automatically by a fuse or circuit breaker if the current becomes too high because of a fault.

2 The thin wire in a fuse melts if the current is too high, which breaks the current.

exam tip >>

Make sure you can recognise wiring errors and dangerous uses of electricity.

>> practice questions

1 Describe the differences between d.c. and a.c. supplies.

2 Why is the case of a plug made from plastic, and its pins from brass?

3 What is the frequency and voltage of the domestic mains supply?

Power

- Power is the rate of transformation of energy.

- The correct fuse for an appliance can be chosen by considering its power and potential difference.

A Power

① Power is a measure of how quickly energy is transformed by a device. It is measured in watts, W.

② **key fact** You can calculate power using this equation:

$$\text{power (W)} = \frac{\text{energy transformed (J)}}{\text{time (s)}}$$

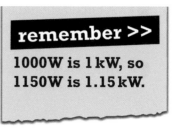

③ *Example 1*
A boy uses 1000J running upstairs in 5s. What was his power?

$$\text{power} = \frac{\text{energy transformed}}{\text{time}} = \frac{1000}{5} = 200\,\text{W}$$

④ *Example 2*
A man lifts a 2.5 N can by 1.2 m in 2 s. What was his power?

work done = force x distance = 2.5 x 1.2 = 3.0 J

$$\text{power} = \frac{\text{energy transformed}}{\text{time}} = \frac{3.0}{2} = 1.5\,\text{W}$$

B Power in electrical circuits

① **key fact** You can calculate power in electrical circuits using this equation:

power (W) = current (A) × potential difference (V)

② For example, a vacuum cleaner draws a current of 5 A at 230V. What is its power?

power = current × potential difference = 5 × 230 = 1150W

remember >>
1000W is 1 kW, so
1150W is 1.15 kW.

C Fuse ratings

1 Fuses come in standard ratings such as 3 A, 5 A and 13 A.

2 **key fact** You should select the fuse with a rating just above the rating of the appliance.

3 For example, an appliance rated at 4 A should be fitted with a 5 A fuse:

- a 3 A fuse would 'blow' all the time and switch off the circuit even when nothing was wrong

- a 13 A fuse would not protect the user of the appliance if a fault happened until the current was dangerously high.

4 **key fact** You can calculate the current flowing through an appliance using this equation:

$$\text{current (A)} = \frac{\text{power (W)}}{\text{potential difference (V)}}$$

5 For example, a 2.3 kW electric heater operates at 230 V. What is the current flowing through it?

$$\text{current} = \frac{\text{power}}{\text{potential difference}} = \frac{2300}{230} = 10\,\text{A}$$

The heater should be protected with a 13 A fuse.

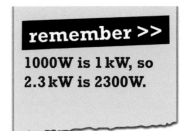

remember >>

1000 W is 1 kW, so 2.3 kW is 2300 W.

>> practice questions

1 An electric lamp transforms 200 J of energy in 2 s. What is its power?

2 A television is rated at 460 W.

   a) What current flows through it at 230 V?

   b) Which fuse, 3 A, 5 A or 13 A would be the best choice?

Energy and charge – *Higher Tier*

> **Charge, current and time are related: if you know two of these quantities you can calculate the third one.**
>
> **Energy, potential difference and charge are also related.**

A Current, charge and time

1 Electrical charge is measured in coulombs, C.

2 **key fact** You can calculate the amount of charge that flows using this equation:

charge (C) = current (A) × time (s)

3 *Example 1*
How much charge flows in 2 minutes at a current of 10A?

2 minutes = 2 × 60 = 120 s

charge = current × time = 10 × 120 = 1200 C

4 *Example 2*
Copper can be purified by electrolysis using copper electrodes and copper(II) sulfate solution.

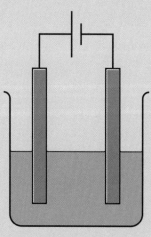

If an electrolysis cell is run for one hour at 2A, how much charge flows?

1 hour = 1 × 60 × 60 = 3600 s

charge = current × time = 2 × 3600 = 7200 C

B Energy and charge

 ① key fact **You can calculate the energy transformed using this equation:**

energy transformed (J) = potential difference (V) × charge (C)

② For example, in the electrolysis cell described earlier, how much energy is transformed if the potential difference applied is 6V?

energy transformed = potential difference × charge
= 6 × 7200 = 43 200 J

C Putting it all together

You know two equations involving charge:

a) charge = current × time

b) energy transformed = potential difference × charge

If you put equation a) into equation b) you get:

c) energy transformed = potential difference × current × time

If you divide both sides by time, you get:

d) energy transformed ÷ time = potential difference × current

Remember from page 88 that power = energy ÷ time, so you get:

e) power = potential difference × current

This is another equation from page 88.

>> practice questions

1 How much charge flows in 30 s at a current of 15 A?

2 How much charge flows in 4 hours at a current of 0.5 A?

3 How much energy is transformed if 20 C of charge flows at a potential difference of 12V?

Radiation

Radioactive decay involving alpha radiation or beta radiation leads to nuclei of different elements being formed.

A   Isotopes and ions

1 Atoms contain protons, neutrons and electrons. These have different masses and charges:

particle	relative mass	relative charge
proton	1	+1
neutron	1	0
electron	almost zero	−1

2 Atoms have no overall electrical charge because the number of protons equals the number of electrons. If atoms gain or lose electrons, they become charged particles called ions.

3 All the atoms of a given element have the same number of protons, and no two elements have the same number of protons. The number of protons in an atom is its atomic number.

4 Isotopes are atoms of an element with different numbers of neutrons. The total number of protons and neutrons in the nucleus of an atom is the mass number.

exam tip >>

This material is also covered in the chemistry examination paper.

B   Effects of alpha decay

1 **key fact** Alpha particles are identical to helium nuclei. They consist of two protons and two neutrons.

2 **key fact** When a radioactive nucleus emits an alpha particle, it loses two protons and two neutrons:

- its atomic number goes down by two

- its mass number goes down by four

- the nucleus of another element is formed.

3 For example, an atom of radium-224 becomes an atom of radon-220 when it emits an alpha particle.

radium-224   radon-220   helium-4
                                         alpha particle

$$^{224}_{88}Ra \rightarrow {}^{220}_{86}Rn + {}^{4}_{2}He$$

C Effects of beta decay

1 **key fact** Beta particles are high-energy electrons given off from the nucleus.

2 **key fact** When a radioactive nucleus emits a beta particle, a neutron becomes a proton:

- **its atomic number goes up by one**

- **its mass number stays the same**

- **the nucleus of another element is formed.**

3 For example, an atom of radium-225 becomes an atom of actinium-225 when it emits a beta particle.

radium-225   actinium-225   beta
                                      particle

$$^{225}_{88}Ra \rightarrow {}^{225}_{89}Ra + {}^{0}_{-1}e$$

> **exam tip >>**
>
> Alpha particles are shown as $^{4}_{2}He$ and beta particles as $^{0}_{-1}e$. Make sure the totals of the top and bottom numbers are the same on both sides of the arrow.

>> practice questions

1 **Explain what happens when a radioactive nucleus emits:**

a) **an alpha particle,**

b) **a beta particle.**

Fusion and fission

- Nuclear reactions involve nuclear fusion or nuclear fission.
- Chain reactions happen in nuclear reactors.
- Background radiation comes from natural and artificial sources.

A Nucleur fusion

1. **key fact** Nuclear fusion is a nuclear reaction in which two nuclei join together to form a larger nucleus.

2. **key fact** Energy is released by stars such as the Sun because of nuclear fusion.

3. For example, in the Sun, hydrogen-1 nuclei fuse with hydrogen-2 nuclei to make helium-3 nuclei:

$$^{1}_{1}H + ^{2}_{1}H \rightarrow ^{3}_{2}He$$

B Nuclear fission

1. **key fact** Nuclear fission is a nuclear reaction in which a nucleus splits into smaller nuclei.

2. Energy is released in nuclear reactors because of nuclear fission.

3. Uranium-235 and plutonium-239 are commonly used in nuclear reactors.

C Chain reactions

1. Nuclear fission reactions in a nuclear reactor start when a uranium-235 nucleus or a plutonium-239 nucleus absorbs a neutron. When this happens:

   - the nucleus splits into two smaller nuclei
   - two or three more neutrons are released
   - energy is released.

2. The newly released neutrons can be absorbed by other uranium-235 nuclei or plutonium-239 nuclei. These nuclei then split, giving off even more neutrons. The reaction is called a chain reaction. It can lead to very many nuclei splitting.

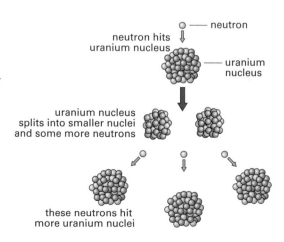

neutron

neutron hits uranium nucleus

uranium nucleus

uranium nucleus splits into smaller nuclei and some more neutrons

these neutrons hit more uranium nuclei

D Background radiation

1 **key fact** **Background radiation is the radiation that is all around us. Some is artificial but most is natural.**

2 Natural radiation includes:

- Cosmic rays (radiation from the Sun and outer space).

- Certain rocks, such as granite, give off radiation. These rocks may be used in buildings. They give off small amounts of radon, a radioactive gas.

- Some foods, such as shellfish and Brazil nuts, contain more radioactive substances than other foods.

3 Artificial background radiation includes:

- radiation received from medical x-rays

- hospital waste and waste from nuclear power stations

- radiation from nuclear weapons tests and accidental releases of radioactive substances.

4 Our bodies are used to these low doses of background radiation. But if the level of radiation becomes too high, there could be problems for your health.

5 You need to take background radiation into account when you are taking measurements of radioactivity.

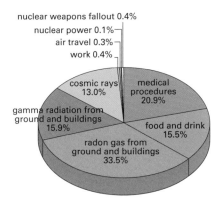

Radiation exposure

>> practice questions

1 **Outline the main difference between nuclear fusion and nuclear fission.**

2 **What is a chain reaction?**

3 **Which source of background radiation, natural or artificial, contributes the most to our exposure to radiation?**

Exam questions

>> Specimen question 1

The diagram shows a plant cell from a leaf.

a) Use the words from the list to label the parts **A** to **D**.

cell membrane   cytoplasm   chloroplast   vacuole   nucleus   cell wall

A _____   B _____

C _____   D _____   **4 marks**

b) Name two parts not found in animal cells.

1 _____   2 _____   **2 marks**

[6 marks in total]

>> Specimen question 2

a) Oxygen moves into cells by diffusion. What is diffusion?

_____

_____ **2 marks**

b) A student carried out an investigation about osmosis. They put two bags of dialysis tubing into 10% sucrose solution and left them for an hour. One bag contained 5% sucrose and the other contained 20% sucrose solution.

10% sucrose solution —————   ————— dialysis tubing

20% sucrose solution —————   ————— 5% sucrose solution

i) What is osmosis?

_____

_____ **3 marks**

ii) Explain what will happen to each bag during the experiment.

5% sucrose bag

_____

_____ **2 marks**

20% sucrose bag

_____

_____ **2 marks**

[9 marks in total]

>> Specimen question 3

a) Complete the equation for photosynthesis:

carbon dioxide + _____ $\xrightarrow{\text{light}}$ oxygen + _____ **2 marks**

b) Name the substance found in plant cells that absorbs light energy for photosynthesis.

_____ **1 mark**

c) Apart from increasing the amount of carbon dioxide in the air and the light intensity, state two ways in which you could increase the rate of photosynthesis of a plant.

1 _____

2 _____ **2 marks**

d) Explain why the amount of carbon dioxide in the air might limit the rate of photosynthesis in very bright light.

_____

_____ **2 marks**

e) Plants need minerals for healthy growth.

i) How do plants get the minerals they need?

_____

_____ **2 marks**

ii) Why do plants need nitrates?

_____ **1 mark**

iii) What happens to plants growing without enough nitrates?

_____ **1 mark**

[11 marks in total]

>> Specimen question 4

Scientists working at sea have discovered part of a marine food web:

phytoplankton → zooplankton → shrimp → small fish → large fish

a) Which is the producer in this food chain?

_____ **1 mark**

The scientists were able to estimate the biomass at each stage in the food chain.

phytoplankton	zooplankton	shrimp	small fish	large fish
1 000 000 kg	100 000 kg	10 000 kg	1000 kg	100 kg

b) Explain why it would be difficult to draw a pyramid of biomass for this food chain.

_____

_____ **2 marks**

c)   Give two reasons why there is less material and energy at each stage in the food chain compared to the one before it.

_____

_____ 2 marks

d)   Suggest why it would be more efficient for food production if we ate the small fish instead of the large fish.

_____

_____ 2 marks

e)   Some scientists have suggested adding large amounts of iron filings to the sea. This would provide the phytoplankton with a mineral that is in short supply in the sea. What would be the likely effect on the mass of large fish if the mass of phytoplankton increased?

_____ 1 mark

[8 marks in total]

>> Specimen question 5

a)   Name the process that happens in your cells that releases energy from oxygen and glucose.

_____ 1 mark

b)   Name two products of this process.

1 _____ 2 _____ 2 marks

c)   Where in the cell is most energy released by this process?

_____ 1 mark

[4 marks in total]

>> Specimen question 6

Complete the sentences about enzymes and digestion by crossing out the incorrect word in each pair.

a)   Enzymes are **proteins / living things** that catalyse reactions.

b)   Proteases convert proteins into **fatty acids / amino acids**.

c)   The enzymes in the stomach work best in **acidic conditions / alkaline conditions**. **3 marks**

>> Specimen question 7

The conditions inside our bodies stay the same.

a)   Complete the table using words from this list:

lungs   kidneys   bladder   liver   brain   skin   stomach 5 marks

	part of the body
makes urine	
stores urine	
produces urea	
produces sweat	
removes carbon dioxide	

b)   Where is the thermoregulatory centre?

_____ **1 mark**

c)   What does the thermoregulatory centre do?

_____ **2 marks**

[8 marks in total]

>> Specimen question 8

a)   Complete the genetic diagram to show how the allele that causes
        Huntington's disease can be inherited. **2 marks**

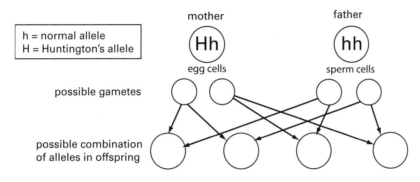

b)   Put a tick against the individuals in the diagram who would
        have Huntington's disease. **2 marks**

c)   Cystic fibrosis is caused by a recessive allele.

   i) Why can people be carriers of the allele and show no signs of the disorder?

   _____

   _____ **2 marks**

   ii) Explain why two parents who are both carriers may produce a child
        who has cystic fibrosis.

   _____

   _____

   _____ **2 marks**

[8 marks in total]

Chemistry

The periodic table and reactivity series on pages 107 and 108 may help you with these questions.

>> Specimen question 1

a) The diagram shows a lithium atom. Use the correct words from the list to label it.   **3 marks**

nucleus   proton   neutron   electron   energy level

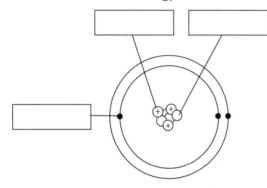

b) Complete this diagram to show the electronic structure of magnesium, Mg.   **2 marks**

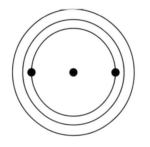

[5 marks in total]

>> Specimen question 2

a) Here is an equation for a reaction:

lead nitrate + potassium iodide → lead iodide + potassium nitrate

What information does it give us?

_____

_____ **2 marks**

b) The formula for one of the substances is $Pb(NO_3)_2$. Name the elements it contains.

_____ **1 mark**

[3 marks in total]

>> Specimen question 3

A student set up an experiment to investigate the rate of reaction of magnesium with an excess of hydrochloric acid. The student used a measuring cylinder to find the total volume of hydrogen produced every 20 seconds. The results are shown in the table to the right.

Time (s)	Volume (cm³)
0	0
20	24
40	36
60	43
80	47
100	48
120	48

a)   Plot the results and draw a curve of best fit. **5 marks**

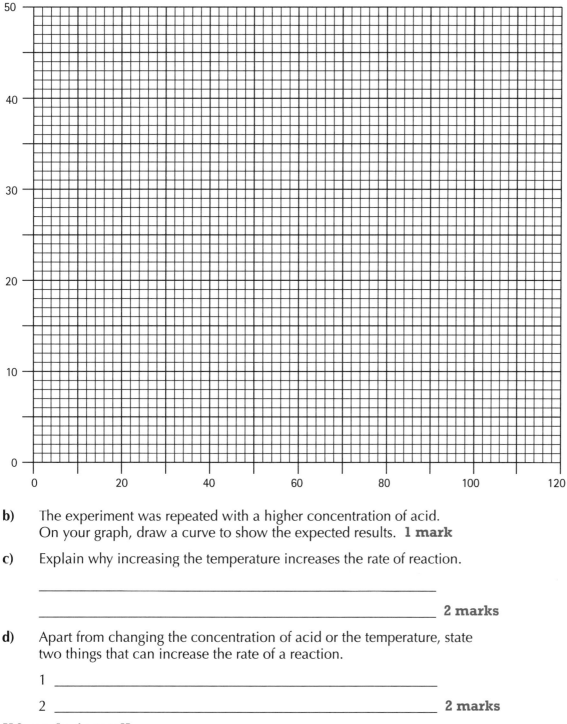

b)   The experiment was repeated with a higher concentration of acid.
On your graph, draw a curve to show the expected results. **1 mark**

c)   Explain why increasing the temperature increases the rate of reaction.

_____

_____ **2 marks**

d)   Apart from changing the concentration of acid or the temperature, state two things that can increase the rate of a reaction.

1 _____

2 _____ **2 marks**

[10 marks in total]

>> Specimen question 4

a) Name the acid that reacts with sodium hydroxide to produce sodium chloride.

_____ 1 mark

b) Write the symbol equation for neutralisation.

_____ 1 mark

c) The formula for sodium hydroxide is NaOH.
Calculate the percentage by mass of oxygen in sodium hydroxide.

_____

_____ 2 marks

[4 marks in total]

>> Specimen question 5

Copper is purified by electrolysis.

a) Which electrode, positive or negative, should be pure copper?

_____ 1 mark

b) Complete the table to show the products expected at each electrode during electrolysis. **4 marks**

substance electrolysed	product at negative electrode	product at positive electrode
molten sodium chloride		
sulfuric acid		
potassium bromide solution		

[5 marks in total]

>> Specimen question 6

Ammonia is used to make fertilisers. The flow chart shows some features of the Haber process, which is used to make ammonia.

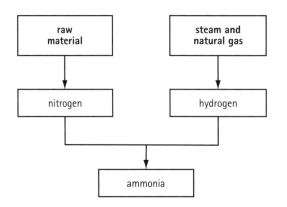

a) Name the raw material from which nitrogen is obtained.

_____ 1 mark

b) Name the catalyst used in the Haber process.

_____ 1 mark

c) **i)** Balance the symbol equation for the Haber process:

$N_2(g) + \ldots H_2(g) \rightleftharpoons \ldots NH_3(g) + heat$ **1 mark**

ii) What does the symbol $\rightleftharpoons$ mean? _____ **1 mark**

d) Explain why the yield of ammonia increases as the pressure increases.

_____ **2 marks**

e) **i)** Explain what happens to the equilibrium yield of ammonia as the temperature is increased.

_____

_____

ii) Suggest one reason why the Haber process is carried out at a moderately high temperature of about 450 °C.

_____

_____ **4 marks**

[10 marks in total]

Physics

>> Specimen question 1

The table shows the velocity of a car at different times from the start of a short journey.

velocity (m/s)	0	25	25	15	8	5	2
time (s)	0	10	20	30	40	50	60

a) Plot the results and draw straight lines of best fit.
One of the points has been plotted for you. **3 marks**

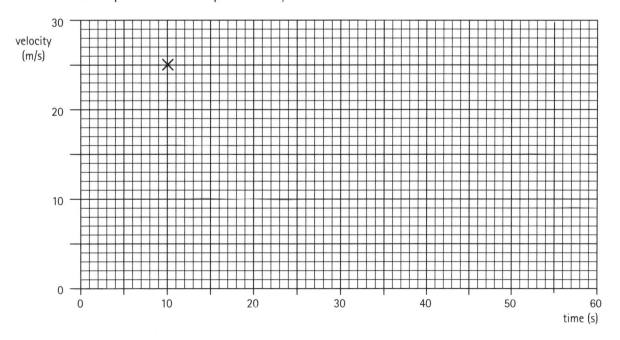

b) Use your graph to answer these questions.

i) Between which two times was the car travelling at a constant velocity?

Between _____ s and _____ s **1 mark**

ii) Between which two times was the car slowing down?

Between _____ s and _____ s **1 mark**

iii) How far did the car travel between 10 s and 20 s?

_____ m **2 marks**

c) Calculate the car's acceleration in the first 10 seconds.

_____

_____ **3 marks**

[10 marks in total]

>> Specimen question 2

A hovercraft is moving across the sea.

forward push

friction

a) In which direction does the resultant force act?

_____ **1 mark**

b) Explain why the hovercraft will eventually reach a terminal velocity.

_____

_____ **2 marks**

c) The hovercraft has a mass of 300 000 kg.

i) How much upwards force must be applied to balance the weight of the hovercraft?

_____

_____ **2 marks**

ii) What is its acceleration if a resultant force of 150 000 N is applied by its propellers?

_____

_____ **2 marks**

d) Calculate the momentum of the hovercraft when it is travelling at 8 m/s.

_____ kg m/s **1 mark**

[8 marks in total]

Higher Tier

e) Calculate the kinetic energy of the hovercraft when it is travelling at 8 m/s.

_____

_____ **2 marks**

[10 marks in total]

>> Specimen question 3

Aircraft are refuelled from tankers. A static electric charge can build up on the tanker as the fuel flows through the pipe to the aircraft.

a) Why is the build up static charge on the tanker dangerous?

_____

_____ 1 mark

b) Explain what the wire connecting the tanker to the ground does.

_____

_____ 1 mark

c) Suggest why the aircraft is also connected to the ground by a wire.

_____

_____ 1 mark

[3 marks in total]

>> Specimen question 4

a) In the space and using a pencil, draw a circuit diagram showing a cell, lamp and open switch connected in series. **3 marks**

b) Explain how you could find the following:

i) The current flowing through the lamp.

_____

_____ 2 marks

ii) The potential difference across the lamp.

_____

_____ 2 marks

c) The potential difference across the lamp is 1.5V and the current flowing through it is 0.5A. What is the resistance of the lamp?

_____

_____ 3 marks

d) Another lamp is connected in parallel to the first lamp. What is the potential difference across this second lamp?

_____ 1 mark

[11 marks in total]

>> Specimen question 5

a) In the space and using a pencil, sketch a current-potential difference graph for a diode and for a filament lamp. **2 marks**

b) Explain how an electrical fuse works.

_____

_____ **2 marks**

c) Write down the equation that links power, current and potential difference.

_____ **1 mark**

d) A computer is connected to the main electricity supply, which is 230V.

i) Calculate the power if 2 A flows through the computer.

_____ **1 mark**

ii) Calculate the amount of energy transformed by the computer in 60 seconds.

_____

_____ **2 marks**

[8 marks in total]

Higher Tier

e) Calculate the charge that flows when the computer is run for 10 minutes.

_____

_____ **2 marks**

[10 marks in total]

>> Specimen question 6

a) Name two sources of natural background radiation.

_____

_____ **2 marks**

b) Uranium-235 is used as a fuel in nuclear power stations.

i) What sort of nuclear reaction takes place?

_____

_____ **1 mark**

ii) Explain how neutrons are involved in chain reactions with uranium-235 nuclei.

_____

_____ **3 marks**

[6 marks in total]

Chemistry data

The periodic table of elements

Key:

relative atomic mass A_r	1	
	H	
	Hydrogen	
atomic number	1	

1	2											3	4	5	6	7	0
																	4 **He** Helium 2
7 **Li** Lithium 3	9 **Be** Beryllium 4											11 **B** Boron 5	12 **C** Carbon 6	14 **N** Nitrogen 7	16 **O** Oxygen 8	19 **F** Fluorine 9	20 **Ne** Neon 10
23 **Na** Sodium 11	24 **Mg** Magnesium 12											27 **Al** Aluminium 13	28 **Si** Silicon 14	31 **P** Phosphorus 15	32 **S** Sulfur 16	35.5 **Cl** Chlorine 17	40 **Ar** Argon 18
39 **K** Potassium 19	40 **Ca** Calcium 20	45 **Sc** Scandium 21	48 **Ti** Titanium 22	51 **V** Vanadium 23	52 **Cr** Chromium 24	55 **Mn** Manganese 25	56 **Fe** Iron 26	59 **Co** Cobalt 27	59 **Ni** Nickel 28	63.5 **Cu** Copper 29	65 **Zn** Zinc 30	70 **Ga** Gallium 31	73 **Ge** Germanium 32	75 **As** Arsenic 33	79 **Se** Selenium 34	80 **Br** Bromine 35	84 **Kr** Krypton 36
85 **Rb** Rubidium 37	88 **Sr** Strontium 38	89 **Y** Yttrium 39	91 **Zr** Zirconium 40	93 **Nb** Niobium 41	96 **Mo** Molybdenum 42	[98] **Tc** Technetium 43	101 **Ru** Ruthenium 44	103 **Rh** Rhodium 45	106 **Pd** Palladium 46	108 **Ag** Silver 47	112 **Cd** Cadmium 48	115 **In** Indium 49	119 **Sn** Tin 50	122 **Sb** Antimony 51	128 **Te** Tellurium 52	127 **I** Iodine 53	131 **Xe** Xenon 54
133 **Cs** Caesium 55	137 **Ba** Barium 56	139 **La** Lanthanum 57	178 **Hf** Hafnium 72	181 **Ta** Tantalum 73	184 **W** Tungsten 74	186 **Re** Rhenium 75	190 **Os** Osmium 76	192 **Ir** Iridium 77	195 **Pt** Platinum 78	197 **Au** Gold 79	201 **Hg** Mercury 80	204 **Tl** Thallium 81	207 **Pb** Lead 82	209 **Bi** Bismuth 83	[209] **Po** Polonium 84	[210] **At** Astatine 85	[222] **Rn** Radon 86
[223] **Fr** Francium 87	[226] **Ra** Radium 88	[227] **Ac** Actinium 89	[261] **Rf** Rutherfordium 104	[262] **Db** Dubnium 105	[266] **Sg** Seaborgium 106	[264] **Bh** Bohrium 107	[277] **Hs** Hassium 108	[268] **Mt** Meitnerium 109	[271] **Ds** Darmstadtium 110	[272] **Rg** Roentgenium 111							

elements 112 to 116 have also been reported

The formulae of some common ions

positive ions			negative ions	
name	formula		name	formula
aluminium	Al^{3+}		bromide	Br^-
ammonium	NH_4^+		carbonate	CO_3^{2-}
barium	Ba^{2+}		chloride	Cl^-
calcium	Ca^{2+}		fluoride	F^-
copper(II)	Cu^{2+}		hydroxide	OH^-
hydrogen	H^+		iodide	I^-
iron(II)	Fe^{2+}		nitrate	NO_3^-
iron(III)	Fe^{3+}		oxide	O^{2-}
lead	Pb^{2+}		sulfate	SO_4^{2-}
lithium	Li^+		sulfide	S^{2-}
magnesium	Mg^{2+}			
potassium	K^+			
silver	Ag^+			
sodium	Na^+			
zinc	Zn^{2+}			

The reactivity series of metals

Carbon and hydrogen are non-metals but are here for comparisons to be made.

most reactive — potassium
sodium
calcium
magnesium
aluminium
carbon
zinc
iron
tin
lead
hydrogen
copper
silver
gold
least reactive — platinum

Answers to exam questions

Biology

1 a) 1 mark for each correct answer:
   A nucleus
   B vacuole
   C chloroplast
   D cell membrane

   b) two from the following, for 1 mark each: cell wall, chloroplast, vacuole

2 a) the movement of particles (1 mark) from a region of high concentration to a region of low concentration (1 mark)

   b) i) Osmosis is the movement of water (1 mark) from a dilute solution to a more concentrated solution (1 mark) through a partially permeable membrane (1 mark).

   ii) 5% sucrose bag: gets smaller/loses water (1 mark), because it is less concentrated than its surroundings (1 mark)

   20% sucrose bag: gets bigger/gains water (1 mark), because it is more concentrated than its surroundings (1 mark)

3 a) in order: water, glucose (1 mark each)

   b) chlorophyll (1 mark)

   c) increase water supply, increase temperature (1 mark each)

   d) The light intensity is high enough for a faster rate (1 mark) but there is not enough carbon dioxide for this rate (1 mark).

   e) i) through their roots / root hair cells (1 mark) dissolved in water (1 mark)

   ii) to make amino acids / to make proteins (1 mark)

   iii) they have poor growth / they are stunted (1 mark)

4 a) phytoplankton (1 mark)

   b) Pyramids of biomass are drawn to scale (1 mark) and the biomasses vary from 100 kg to 1 000 000 kg, which is a very large range (1 mark).

   c) any two from the following for 1 mark each: energy is lost through respiration, energy is lost through movement, energy is lost in waste materials, materials are lost in waste

   d) The food chain would be shorter / the biomass of small fish is much more than the biomass of large fish (2 marks).

e) it would increase (1 mark)

5 a) (aerobic) respiration (1 mark)

   b) water, carbon dioxide (1 mark each, either order)

   c) the mitochondria (1 mark)

6 a) proteins (1 mark)

   b) amino acids (1 mark)

   c) acidic conditions (1 mark)

7 a) 1 mark for each correct line in the table

	part of the body
makes urine	kidneys
stores urine	bladder
produces urea	liver
produces sweat	skin
removes carbon dioxide	lungs

   b) brain (1 mark)

   c) monitors (1 mark) and controls body temperature (1 mark)

8 a) completed diagram, 1 mark for each correct row to 2 marks

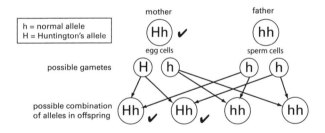

   b) mother ticked (1 mark), two offspring on the left ticked (1 mark)

   c) i) Two copies are needed to have the disorder (1 mark) and carriers only have one copy (1 mark).

   ii) If both parents are carriers they each have one copy of the allele (1 mark), so there is the possibility of producing a child who has two copies (1 mark).

Chemistry

1 a) 1 mark for each correct label.

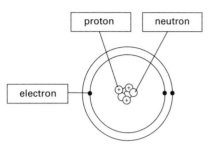

b) 1 mark for 12 electrons, 1 mark for correct number in each circle (2, 8 and 2)

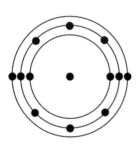

2 a) 1 mark for each of: lead nitrate and potassium iodide are reactants; lead iodide and potassium nitrate are products

b) lead, nitrogen, oxygen (all three correct for 1 mark)

3 a) 1 mark for labelling axes with units; 1 mark for smooth curve of best fit; 3 marks for correct plots (take away 1 mark for each incorrect point until no marks for plotting)

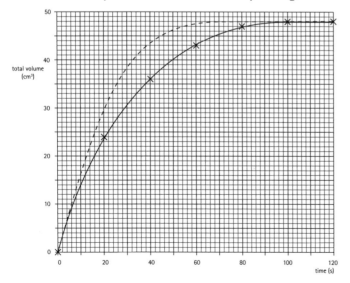

b) a line steeper and to the left of the original, finishing at the same height for 1 mark

c) two from the following, 1 mark each: particles move faster; more chance of colliding; more particles have the activation energy

d) 1 mark each: add a catalyst, powder the magnesium

4 a) hydrochloric acid (1 mark)

b) $H^+(aq) + OH^-(aq) \rightarrow H_2O(l)$ (1 mark)

c) M_r of NaOH = 23 + 16 + 1 = 40 (1 mark)

% O = 16 ÷ 40 × 100 = 40% (1 mark)

5 a) negative (1 mark)

b) 1 mark for each correct answer to max. 4 marks:

substance electrolysed	product at negative electrode	product at positive electrode
molten sodium chloride	sodium	chlorine
sulfuric acid	hydrogen	oxygen
potassium bromide solution	hydrogen	bromine

6 a) air (1 mark)

b) iron (1 mark)

c) i) $N_2(g) + 3H_2(g) \rightleftharpoons 2NH_3(g)$ + heat (1 mark)

ii) reversible reaction (1 mark)

d) There are more molecules of gas on the right (1 mark) so as the pressure increases the position of equilibrium moves to the right (1 mark).

e) i) forward reaction is exothermic (1 mark) so the position of equilibrium moves to the left as the temperature is increased (1 mark)

ii) temperature needs to be high enough to get a good rate of reaction (1 mark) but not so high that the yield is low (1 mark)

Physics

1 a) 2 marks for correct plots (take away 1 mark for each incorrect point until no marks for plotting); 1 mark for lines of best fit.

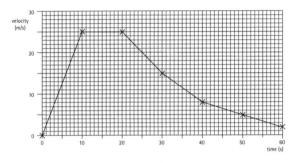

   b) i) between 10 s and 20 s (1 mark)

      ii) between 20 s and 60 s (1 mark)

      iii) $(20 - 10) \times 25 = 10 \times 25 = 250$ m (1 mark for working, 1 mark for answer)

   c) $(25 - 0) \div (10 - 0) = 25 \div 10 = 2.5$ m/s^2
      1 mark for working, 1 mark for answer,
      1 mark for correct units

2 a) forwards / to the left (1 mark)

   b) The friction will increase as the hovercraft goes faster (1 mark) until it balances the forward push (1 mark).

   c) i) $300\,000 \times 10 = 3\,000\,000$ N (1 mark for working, 1 mark for answer)

      ii) $150\,000 \div 300\,000 = 0.5$ m/s^2 (1 mark for answer, 1 mark for correct units)

   d) $300\,000 \times 8 = 2\,400\,000$ kg m/s (1 mark)

   e) $\frac{1}{2} \times 300\,000 \times 8^2 = 9\,600\,000$ J (1 mark for working, 1 mark for answer)

3 a) A spark could ignite the fuel / cause an explosion (1 mark).

   b) It discharges the electric charge to earth (1 mark).

   c) The aircraft also becomes charged so it needs to be earthed (1 mark).

4 a) 1 mark for each correct circuit symbol up to 2 marks; 1 mark for showing them in series (any order).

   b) i) Use an ammeter (1 mark) connected in series (1 mark).

      ii) Use a voltmeter (1 mark) connected in parallel (1 mark).

   c) $1.5 \div 0.5 = 3.0\,\Omega$ (1 mark for working, 1 mark for answer, 1 mark for correct units)

   d) the same / 1.5 V (1 mark)

5 a) 1 mark for each correct graph

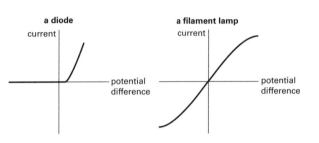

   b) A wire in the fuse melts and breaks the circuit (1 mark) if the current gets too high (1 mark).

   c) power = current × potential difference (1 mark)

   d) i) $2 \times 230 = 460$ W (1 mark for answer, 1 mark for correct unit)

      ii) $460 \times 60 = 27\,600$ J (1 mark for working, 1 mark for answer)

   e) $2 \times (10 \times 60) = 1200$ C (1 mark for answer, 1 mark for correct unit)

6 a) any two of the following for 1 mark each: cosmic rays, radon, food, drink, granite

   b) i) nuclear fission (1 mark)

      ii) A neutron hits a uranium-235 nucleus causing it to split (1 mark) which releases 2 or 3 more neutrons (1 mark) which go on to split even more uranium-235 nuclei (1 mark).

Answers to practice questions

Biology

Cells (p3)

1 a) cell membrane, nucleus, cytoplasm, mitochondria, ribosomes

   b) cell wall, chloroplasts, permanent vacuole

2 The permanent vacuole is not filled with cell sap so the cytoplasm is not pushed against the cell wall.

3 enzymes

Diffusion and osmosis (p5)

1 Diffusion is the movement of a substance from a region where it is more concentrated to a region where it is less concentrated.

2 The fruit cells lost water by osmosis. The solution inside the cell was more dilute than the solution outside the cell so the water moved from the dilute solution inside the cell, through the semi-permeable cell membrane, to the more concentrated solution outside the cell.

Photosynthesis (p7)

1 carbon dioxide + water (+ light energy) → glucose + oxygen

2 respiration; converted to starch for storage

3 At a particular light intensity, the amount of oxygen being produced in photosynthesis will be the same as that used up in respiration.

Minerals (p9)

1 Plants absorb dissolved minerals through their roots.

2 Nitrate ions are needed to make amino acids for proteins. Magnesium ions are needed to make chlorophyll. Nitrate deficiency causes stunted growth and magnesium deficiency causes yellow leaves.

Food chains and pyramids (p11)

1 the Sun

2 A primary consumer eats plants but a secondary consumer eats animals.

3 The biomass at each stage of a food chain. The width of each bar in the pyramid represents the amount of biomass.

Food production (p13)

1 Some energy is lost in the production of waste materials and released by respiration.

2 Much more energy is wasted in meat production than in cereal production. This evidence supports the view that more people could be fed if everyone became vegetarian.

3 Less energy is lost to the environment if movement is restricted. Less energy is lost in maintaining body temperature if the chickens are kept in a warm environment.

The carbon cycle (p15)

1 photosynthesis

2 respiration

3 More carbon dioxide will be available for photosynthesis.

Respiration and enzymes (p17)

1 a biological catalyst

2 respiration, photosynthesis, protein synthesis from amino acids

3 They become denatured. Their shape changes so much that they cannot catalyse the reaction any more.

Digestion (p19)

1 mouth and small intestine

2 Lipases catalyse the breakdown of lipids (fats and oils) into fatty acids and glycerol.

3 The proteases in the stomach work best in acidic conditions but the proteases in the small intestine work best in alkaline conditions.

Uses of microorganisms and enzymes (p21)

1 They contain lipases, to digest the fat and oils in stains, and proteases to digest the proteins in stains. These products are soluble in water and are more easily removed during washing.

2 less energy needed to warm the water

3 Carbohydrase converts starch syrup into sugar syrup. Starch syrup is not sweet but is relatively inexpensive, but sugar syrup is sweet and more expensive. It is often used in sports drinks. Isomerase converts glucose syrup into fructose syrup. Fructose syrup is sweeter than the same mass of glucose syrup. It is used in slimming foods.

Maintaining internal conditions (p23)

1 ion content, water content, temperature, blood sugar concentration

2 Urea is produced in the liver from excess amino acids. It is removed from the body in urine by the kidneys.

3 The thermoregulatory centre controls and monitors body temperature.

Diabetes (p25)

1 the pancreas

2 Diabetes is a condition where blood glucose concentration is not controlled sufficiently by hormones and can increase to a level that is fatal.

3 Insulin reduces the blood glucose concentration if that becomes too high. It allows glucose to move from the blood into the cells, and it causes the liver to convert glucose into glycogen, which is stored.

DNA (p27)

1 DNA is deoxyribose nucleic acid. It is found in the nucleus of a cell.

2 23 pairs of chromosomes

3 Men have two different sex chromosomes, XY, and women have the same two sex chromosomes, XX.

Cell division (p29)

1 They contain the same genes.

2 When an embryo forms, half of each pair of alleles comes from one parent, and half of each pair comes from the other parent.

3 An allele is a particular form of a gene.

Inheritance (p31)

1 The F allele is the dominant allele. Any plant containing an F allele will have red flowers, so FF and Ff plants will both have red flowers. A plant containing two recessive ff alleles will have white flowers.

Inherited disorders (p33)

1 It is caused by a dominant allele so only one copy of the allele is needed to cause Huntington's disease.

2 An individual with one copy of a recessive allele which is responsible for causing a genetic disorder.

3 It is caused by a recessive allele, so two copies must be present to have cystic fibrosis.

Inheritance – Higher Tier (p35)

1 The chromosomes are copied, the cell divides twice, and four gametes are formed, each with a single set of chromosomes.

2 a) 25% BB, 50% Bb, 25% bb

b) 25% of the offspring will have blue eyes

Chemistry

Atomic structure (p37)

1

element	configuration
H	1
He	2
Li	2,1
Be	2,2
B	2,3
C	2,4
N	2,5
O	2,6
F	2,7
Ne	2,8
Na	2,8,1
Mg	2,8,2
Al	2,8,3
Si	2,8,4
P	2,8,5
S	2,8,6
Cl	2,8,7
Ar	2,8,8
K	2,8,8,1
Ca	2,8,8,2

Ions and ionic bonds (p39)

1 a)

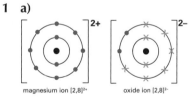

magnesium ion [2,8]$^{2+}$ oxide ion [2,8]$^{2-}$

b)

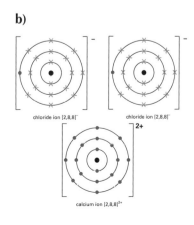

chloride ion [2,8,8]$^{-}$ chloride ion [2,8,8]$^{-}$

calcium ion [2,8,8]$^{2+}$

Simple molecules (p41)

1 a shared pair of electrons

2

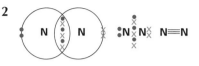

Structure and properties of materials (p43)

1 Graphite consists of layers of carbon atoms with weak forces between the layers. The layers can slide over each other, so graphite is soft and slippery.

2 The ions are free to move in aqueous sodium chloride but not in solid sodium chloride.

3 Giant covalent molecules (macromolecules), giant ionic lattices and metals have high melting points.

Structure and bonding – Higher Tier (p45)

1 Metals have giant structures in which atoms are regularly arranged. Electrons in the highest energy level are delocalised. Metallic bonds are the electrostatic forces of attraction between positively charged ions and delocalised electrons.

2 Metals and graphite have electrons that are free to move.

3 They have weak forces of attraction between the molecules which need little energy to overcome.

Relative atomic mass (p47)

1 atoms of the same element that have different numbers of neutrons

2 seven times heavier

3 a) $40 + 16 = 56$

b) $1 + 1 + 32 = 34$

c) $40 + 32 + 16 + 16 + 16 + 16 = 136$

d) $40 + 16 + 1 + 16 + 1 = 74$

Formulae and percentage composition (p49)

1 a) CaO

   b) Fe_2S_3

   c) $Al(OH)_3$

2 $16 \div 40 \times 100 = 40\%$

Atom economy and balancing equations (p51)

1 Making sure we can meet our own needs without harming the ability of future generations to meet their needs.

2 They produce less waste than reactions with low atom economies, are cheaper to run and use less of the Earth's resources.

3 $2Na(s) + 2H_2O(l) \rightarrow$
   $2NaOH(aq) + H_2(g)$

Chemical calculations – Higher Tier (p53)

1 a) $\frac{1.2}{24} \times (24 + 16) = 2.0\,g$

   b) $\frac{40}{(40 + 63.5)} \times 100 = 38.6\%$

2 SO_2

Rates of reaction (p55)

1 Reacting particles must collide with each other with energy greater than or equal to the activation energy of the reaction.

2 a) When the temperature is increased, the reactant particles move more quickly and more of them have the activation energy. If the rate of successful collisions increases, so does the rate of the reaction.

   b) When the surface area is increased, more solid reactant particles are exposed. If the rate of successful collisions increases, so does the rate of the reaction.

   c) When the concentration of dissolved reactants increases, there are more reacting particles in a given volume of solution. There is a greater chance of a successful collision. If the rate of successful collisions increases, so does the rate of the reaction.

Reversible reactions (p57)

1 Heat energy will be given out because it is an exothermic reaction.

2 The reverse reaction will be endothermic.

The Haber process (p59)

1 Artificial fertilisers are made from ammonia, which is made from nitrogen and hydrogen. The hydrogen is made by reacting natural gas or coal (fossil fuels) with steam.

2 The Haber process uses a high temperature, a high pressure and an iron catalyst.

Electrolysis (p61)

1 a) Hydrogen is released from the cathode (negative electrode).

   Oxygen is released from the anode (positive electrode).

   b) Hydrogen ions are reduced when they reach the electrode because they gain electrons.

2 a) hydrogen gas

   b) copper

Acids, bases and neutralisation (p63)

1 Alkalis are soluble bases. Not all bases are soluble, so therefore not all bases can be alkalis.

2 a) Acids release hydrogen ions, H^+

   Bases release hydroxide ions, OH^-

   b) Hydrogen ions react with hydroxide ions to form water:
   $H^+ + OH^- \rightarrow H_2O$

Making salts (p65)

1 lead nitrate and any one from: sodium iodide, potassium iodide or ammonium iodide

2 Potassium is a very reactive metal and so it would be very dangerous to react it with nitric acid.

Physics

Speed, distance and acceleration (p67)

1 $100 \div 20 = 5\,m/s$

2 $(30 - 15) \div 30 = 0.5\,m/s^2$

3 $50 \times 20 = 1000\,m$

Forces (p69)

1 a) The resultant force is not zero.

   b) The lorry will accelerate forwards.

Weight and falling (p71)

1 a) As the parachutist falls her weight stays the same but the air resistance increases. Eventually her weight is balanced by the air resistance and the resultant force becomes zero.

   b) Her speed increases until she reaches terminal velocity when her weight is balanced by air resistance.

Stopping distances (p73)

1 a) It will increase because the braking distance and thinking distance increase the faster the car travels.

   b) It will increase because worn brakes will produce a weaker braking force.

   c) It will increase because there will be less friction between the road and the tyres.

2 a) $15 \times 0.6 = 9\,m$

   b) $9 + 16.7 = 25.7\,m$

Work and kinetic energy (p75

1 $1.5 \times 2 = 3\,J$

2 A golf ball, because it has a greater mass than a table tennis ball.

3 A beach ball can store elastic potential energy because its shape can be changed by squashing. A stone cannot be squashed and so it cannot store elastic potential energy.

4 kinetic energy =
   $\frac{1}{2} \times 10 \times 4^2 = 80\,J$

Momentum (p77)

1 Momentum =
0.06 × 50 = 3 kg m/s

2 Momentum of person =
50 × 1 = 50 kg m/s

Speed of boat =
50 ÷ 100 = 0.5 m/s

3 Force needed = 25 ÷ 5 = 5 N

Static electricity (p79)

1 The balloon becomes electrically charged because electrons have been transferred. It attracts the hair because the balloon and the hair have opposite electrical charges.

2 Fuel passing along a refuelling pipe causes enough friction to give the pipe an electrostatic charge. If the charge builds up there is the danger of a spark that may cause an explosion. By connecting the pipe to earth with a wire the pipe can be discharged safely.

Electrical circuits (p81)

1

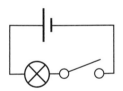

2 Current is measured with an ammeter in series with the component. Potential difference is measured with a voltmeter in parallel with the component.

Resistance (p83)

1 Resistance = 250 ÷ 5 = 50 Ω

2 a) straight line with a positive gradient

b) an 'S' shaped curved

c) No current when the potential difference is negative, then increasing rapidly to a straight line with a positive gradient.

Parallel and series circuits (p85)

1 a) potential difference of one cell
= 6 ÷ 3 = 2V

b) 6V

c) 4V

Mains electricity (p87)

1 The current passes in the same direction in d.c. but it changes direction constantly in a.c.

2 Plastic is an insulator so you do not get an electric shock. Brass is a good conductor of electricity so current can pass easily through the pins.

3 50 Hz, 230V

Power (p89)

1 power = 200 ÷ 2 = 100W

2 a) current = 460 ÷ 230 = 2A

b) 3A

Energy and charge – Higher Tier (p91)

1 charge = 15 × 30 = 450 C

2 charge =
4 × 60 × 60 × 0.5 = 7200 C

3 energy = 12 × 20 = 240 J

Radiation (p93)

1 a) The atomic number decreases by 2 and the mass number decreases by 2, because two protons and two neutrons have been lost.

b) The atomic number increases by 1 but the mass number does not change.

Fusion and fission (p95)

1 Nuclear fusion: two smaller nuclei join to make a larger nucleus.

Nuclear fission: a large nucleus breaks apart to form smaller nuclei.

2 A process in which the products cause more processes of the same kind to happen.

3 natural background radiation

Glossary

accelerate To change speed

activation energy The minimum energy needed for a collision between reactant particles to cause a reaction.

aerobic respiration Reaction needing oxygen which releases energy in cells

air resistance Frictional force acting on objects moving through the air

alkali A soluble base. Alkalis release hydroxide ions when added to water and have a pH greater than 7.

alleles Different forms of a gene

alternating current An electric current which regularly changes direction, abbreviated to a.c.

ammeter Device for measuring electrical current

ampere The unit of electrical current, symbol A

anode Positively charged electrode

asexual reproduction Reproduction needing only one parent, and which produces genetically identical offspring

atomic number The number of protons in the nucleus of an atom, also called the proton number

B

background radiation Radiation that is around us all the time

balanced forces Equal and opposite forces acting on an object

base Metal oxides, hydroxides and carbonates. They neutralise acids. Soluble bases are called alkalis.

bile Acidic liquid produced by the liver, stored in the gall bladder and released into the small intestine

biomass The mass of living material at a particular stage in a food chain

brain The main organ of the central nervous system, responsible for coordinating responses

braking distance The distance travelled by a vehicle between the brakes being applied and the vehicle coming to a halt

C

carrier An individual with an allele for a genetic disorder but who does not have the disorder themselves

catalyst A substance that speeds up chemical reactions without being used up itself

cathode Negatively charged electrode

cell wall Tough layer surrounding the cell membrane in a plant cell

chloroplast Object in the cytoplasm of plant cells where photosynthesis happens

chromosomes Objects in the nucleus carrying genes

circuit breaker An automatic electrical switch that protects a circuit and connected devices if a fault happens

circuit diagram Diagram representing an electrical circuit in symbols and lines

circuit symbol Standard sign for an electrical component, used in circuit diagrams

concentration gradient A difference in concentration that can let diffusion happen

coulomb The unit of electrical charge, symbol C

covalent bond A chemical bond in which a pair of electrons is shared between two atoms

crystallisation Evaporating the water from a solution to leave a solid behind

current Flow of electrical charge

cystic fibrosis A genetic disorder caused by a recessive allele in which thick, sticky mucus is produced in the lungs and gut

D

decay microorganisms Bacteria that break down waste materials and the remains of dead plants and animals

deficiency disease Illness caused by a lack of an important substance

delocalised Delocalised electrons are not associated with any particular atom and are free to move.

denatured When an enzyme's structure is changed so much that the enzyme no longer works

detritus feeders Living things that feed on bits of partly broken down animal and plant material

diabetes Illness in which blood glucose levels can become too high

diffusion The movement of a substance from a region where it is more concentrated to a region where it is less concentrated

digestion Breaking down food so that small, soluble molecules are produced

digestive enzymes Biological catalysts that speed up digestion

direct current An electric current which does not change direction, abbreviated to d.c.

discharge When static electricity leaves a charged object

DNA Complex molecule found in chromosomes which carries the genetic code.

DNA fingerprinting Method to identify individuals from their unique pattern of DNA

dominant allele A form of a gene that only needs to be present as one copy for it to control a particular characteristic

E

earth In electricity, zero electrical potential

elastic potential energy Stored energy because of a change in shape of an object

electrode A rod of conducting material, usually metal or graphite, that is used in electrolysis

electrolysis Using electricity to break down a compound into simpler substances

electron Particle in an atom arranged around its nucleus

electronic structure The arrangement of electrons in energy levels around the nucleus

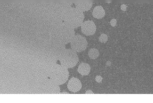

electrostatic charge   Electric charge that builds up on an insulated object

embryo   The early stages of a new life, formed when two gametes fuse

embryo screening   Checking an embryo to see if it carries the alleles for a genetic disorder

energy level   A particular amount of energy an electron around a nucleus can have

enzyme   A protein that can catalyse a chemical reaction

F

fertilisation   Joining or fusion of gametes

fertiliser   Chemical that replaces the minerals in the soil used up by growing plants

filtration   Separating an insoluble solid from a mixture with a liquid using filter paper

food chain   The pathway followed by nutrients and energy in a community

food web   Interconnected food chains

fuse   In electricity, a piece of wire that melts and breaks the circuit if a fault happens. In general, it means joining together.

G

gall bladder   A small sac that stores bile made by the liver

gametes   Sex cells. Sperm are male sex cells and eggs are female sex cells.

gene   Section of DNA in a chromosome that controls a characteristic in the body

genetic   To do with genes and inheritance

genetic diagrams   A chart that shows the possible outcomes of a particular cross

genetic disorder   Disease that is inherited rather than caused by a pathogen

giant covalent structures   A network of atoms joined by many covalent bonds

giant ionic lattices   A regular structure consisting of oppositely charged ions

giant structure   A structure consisting of many atoms or ions

gravitational field strength   The strength of gravity in a particular area. On Earth it is about 10 N/kg.

H

Haber process   The industrial method of making ammonia from nitrogen and hydrogen with an iron catalyst

habitat   The place where an organism lives

hormone   Chemical produced by a gland and transported in the bloodstream to a target organ

Huntington's disease   Genetic disorder caused by a dominant allele. It affects the nervous system in older people.

I

insoluble   Does not dissolve

insulin   A hormone that controls blood glucose concentration

intermolecular forces   Weak forces that attract molecules towards each other

ion   Charged particle made when an atom loses (or gains) an electron

ionic bonding   Electrostatic force of attraction between oppositely-charged ions

isotopes   Atoms with the same number of protons but different numbers of neutrons

J

joule   The unit of energy, symbol J

K

kidney   One of a pair of organs that removes urea, and excess ions and water from the blood

kinetic energy   Energy that any moving object has

L

liver   Organ that has many functions, including making bile and converting excess amino acids into urea

M

macromolecule   Molecule made of very many atoms joined together by covalent bonds

mass number   The number of protons and neutrons in the nucleus of an atom

meiosis   Cell division that produces gametes, which have only one set of chromosomes

metallic bonds   Forces that attract positive metal ions to delocalised electrons in metals

mitosis   Cell division in body cells that produces two identical cells

mole   The unit of amount of substance. The relative atomic mass or relative formula mass of a substance in grams contains one mole.

momentum   A measure of the tendency of an object to keep moving in a certain direction, found by multiplying its mass and velocity together

N

nanoparticles   Particles between 1 nm and 100 nm in size

neutron   Neutral particle found in the nucleus of an atom

nuclear fission   Change to an atom in which its nucleus breaks apart

nuclear fusion   Reaction in which two smaller nuclei join together to make a larger nucleus

nucleus   The part at the centre of the atom

O

ohm The unit of electrical resistance, symbol Ω

osmosis The diffusion of water from a dilute solution to a more concentrated solution through a partially permeable membrane

ovary Part of the female reproduction system where eggs are produced

oxidised A substance that gains oxygen or loses electrons is said to be oxidised.

P

pancreas Gland that secretes digestive enzymes and insulin

parallel A parallel electrical circuit is one in which there is more than one path for the electric current to flow.

partially permeable membrane Thin layer that only allows particles of a certain size to pass through

photosynthesis Process used by green plants to make glucose and oxygen from water and carbon dioxide using light energy

potential difference The change in energy possessed by electrons when they flow through a component, also called the voltage

power The rate of energy transfer

precipitate Insoluble solid formed by reacting two solutions together

precipitation Reaction in which an insoluble solid is formed by reacting two solutions together

proton Positively charged particle in the nucleus

proton number The number of protons in the nucleus of an atom, also called the atomic number

pyramid of biomass Bar chart that shows the mass of living material at each stage in a food chain

R

raw materials The starting substances for a process

receptor Group of cells that detects a stimulus

recessive allele A form of a gene that must be present as two copies for it to control a particular characteristic

reduced A substance that loses oxygen or gains electrons is said to be reduced.

relative atomic mass The mean mass of an atom compared to $\frac{1}{12}$ the mass of a carbon-12 atom. It has the symbol A_r.

relative formula mass The sum of all the relative atomic masses of all the atoms in a substance. It has the symbol M_r.

resistance A measure of how difficult it is for an electric current to pass through a component in a circuit

resistor A component intended to cause resistance in a circuit

respiration The breakdown of glucose in cells to produce water and carbon dioxide, with the release of energy

resultant force The overall force acting on an object, found by adding all the individual forces together and taking into account their direction

S

series A series electrical circuit is one in which there is only one path for the electric current to flow.

sex chromosomes The pair of chromosomes in the nucleus responsible for determining gender. Males have XY chromosomes and females have XX chromosomes.

sexual reproduction Reproduction in which a male and female gamete fuse. It produces genetic variation in the offspring.

shiver Rapid small muscle movements which release heat energy

simple molecule A molecule made of only a few atoms joined together by covalent bonds

small intestine The part of the gut where most digestion and absorption of nutrients happens

speed The change of distance of an object with time

sperm The male gamete

stomach Part of the gut where food travels to after being swallowed

stopping distance The overall distance needed to stop a moving vehicle, including the thinking distance and braking distance

T

terminal velocity The maximum speed reached by a falling object, when its weight is balanced by frictional forces such as air resistance

testes The male sex organs where sperm are produced

thermoregulatory centre The part of the brain that monitors and controls body temperature

thinking distance The distance travelled by a vehicle during the driver's reaction time

U

unbalanced forces When the resultant force on an object is not zero, the forces acting on it are unbalanced.

urea Substance made in the liver by the breakdown of excess amino acids

V

vacuole Space in a plant cell where cell sap is found

variable resistor An electrical component which can change its resistance

velocity The speed of an object in a certain direction

volt The unit of potential difference, symbol V

voltmeter Device used to measure potential difference or voltage

W

watt The unit of power, symbol W

weight The force on an object because of its mass in a gravitational field

work The amount of energy transferred, found by multiplying together the force applied and distance moved

Biology

Cells

- Most animal cells and human cells have these parts:

part	notes
cell membrane	controls the movement of substances into and out of the cell
cytoplasm	most of the cell's chemical reactions take place here
nucleus	controls the cell's activities
mitochondria	where most energy is released by respiration
ribosomes	where protein synthesis happens

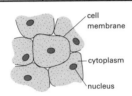

- Plant cells also have these parts:

part	notes
cell wall	strengthens the cell
chloroplasts	absorb light energy to make food by photosynthesis
permanent vacuole	filled with cell sap

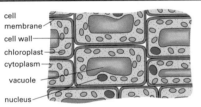

- Diffusion is the spreading of gas particles, or dissolved particles, from high concentration to low concentration. Diffusion happens faster when there is a big concentration gradient. Oxygen for respiration diffuses into cells through the cell membrane.
- Osmosis is the movement of water across a partially permeable membrane from a dilute solution to a more concentrated solution. If the concentration outside a cell is greater than the concentration inside the cell, water moves out of the cell by osmosis.

Plants

- During photosynthesis, carbon dioxide and water react together to make glucose and oxygen:
  carbon dioxide + water (+ light energy) →
  glucose + oxygen
- Some of the glucose is used for respiration. Some is converted into starch. This is insoluble and used for storage.
- Chloroplasts are found in some plant cells. They contain chlorophyll. This is a green substance that absorbs the light energy needed for photosynthesis to work.

- Several factors may limit the rate of photosynthesis, including light intensity, carbon dioxide concentration and temperature.
- Plant roots absorb mineral salts needed for healthy growth.

mineral ion	nitrate	magnesium
why needed	making amino acids for proteins	making chlorophyll
deficiency symptoms	stunted growth	yellow leaves

Chains, webs and food production

- Food chains show the feeding relationships between organisms. Food webs consist of several interconnected food chains.
- Most food webs start with green plants because these capture light energy from the Sun to make food by photosynthesis.
- Biomass is the mass of living material.
- The biomass becomes less at each stage in a food chain.
- Pyramids of biomass are charts drawn to scale. They show the biomass at each stage. They get smaller towards the top.
- The efficiency of food production can be improved by reducing the number of stages in the food chain, and by keeping animals warm and preventing them from moving around. There are ethical and moral considerations to take into account when increasing food production involving animals.

The carbon cycle

- Carbon is recycled in the environment through various compounds.
- Green plants remove carbon from the environment. They absorb carbon dioxide from the atmosphere for photosynthesis.
- Carbon compounds in plants become fats and proteins in animals when they eat the plants.
- Respiration by plants, animals and microorganisms returns carbon to the environment as carbon dioxide.
- Various microorganisms and detritus feeders break down carbon compounds from the remains of dead animals and plants.

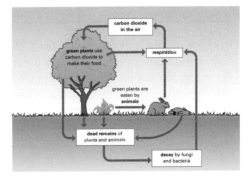

Aerobic respiration
- Aerobic respiration needs oxygen. It is a chemical process in which energy is released from glucose:
  glucose + oxygen → carbon dioxide + water (+ energy)
- Most of the reactions involved in aerobic respiration happen in the mitochondria in cells.
- The energy released by respiration is used by mammals and birds to keep warm. It is used by plants to make amino acids from sugars and nitrates; these are used to make proteins. It is also used to let muscles contract in animals.

Enzymes
- Enzymes are biological catalysts. They are proteins with special shapes that let them speed up chemical reactions.
- Some enzymes are used inside the cell.
  For example, the enzymes involved in respiration.
- Some enzymes are used outside the cell.
  For example, the enzymes involved in digestion.
- Enzymes have household uses in 'biological' detergents. These contain proteases to digest proteins in stains, and lipases to digest fats in stains.
- Enzymes have industrial uses, too.

enzyme	industrial use
carbohydrase	Converts starch into more valuable sugar syrup
isomerase	Converts glucose syrup into sweeter fructose syrup, used in slimming food
protease	Digests the proteins in baby food

Digestion
- Digestion involves the breakdown of large molecules into smaller molecules in the gut. Enzymes are involved in digestion.

enzyme	where produced	reaction catalysed
amylase	salivary glands, pancreas, small intestine	starch → sugars
protease	stomach, pancreas, small intestine	proteins → amino acids
lipase	pancreas, small intestine	lipids → fatty acids + glycerol

- The stomach produces hydrochloric acid. The optimum conditions for enzymes there are acidic.
- The optimum conditions for enzymes in the small intestine are alkaline. The liver produces bile. This is stored in the gall bladder and then released into the small intestine. It neutralises the acid added in the stomach.

Maintaining internal conditions
- Water and ion content, body temperature and blood sugar concentration are all controlled by the body.
- Excess water and ions are removed from the body in the sweat and urine.
- Sweating and making more blood flow through the skin capillaries helps to cool the body down.
- Insulin is a hormone produced by the pancreas. It reduces the blood glucose concentration by causing glucose to move from the blood into the cells.
- Diabetes is a disease in which blood glucose concentrations rise to dangerously high levels. One cause is insufficient insulin production by the pancreas.
- Diabetes can be controlled using injections of insulin and careful diets.

Cell division and DNA
- Sexual reproduction needs two parents. Fertilisation is the fusion of two gametes to form a new cell. It leads to variation.
- Asexual reproduction needs just one parent. The offspring are genetically identical. They are clones.
- DNA is a complex molecule that carries the genetic code. A gene is a small section of DNA that controls a particular characteristic in an organism.
- Each person's DNA is unique, unless they are one of a pair of identical twins. DNA fingerprinting relies on this to indentify individuals.
- Chromosomes are made of DNA and contain many genes. They are found in the nucleus of the cell. Human body cells each contain 23 pairs of chromosomes.
- Females have XX chromosomes and males have XY chromosomes.
- Body cells are produced by mitosis. Gametes are produced by meiosis.

Inheritance
- Genes have different forms called alleles.
- Some alleles are dominant. They control a characteristic even if they are present on just one chromosome. Huntington's disease is an inherited disorder caused by a dominant allele. It can be passed on by just one parent with the disorder.
- Some alleles are recessive. They only control a characteristic if they are present on both chromosomes in a pair. Cystic fibrosis is an inherited disorder caused by a recessive allele. It can be passed on by parents who do not have the disorder themselves.
- Genetic diagrams allow the outcomes of particular crosses to be predicted.

Chemistry

Atomic structure

- Atoms have a central nucleus containing protons and neutrons. The nucleus is surrounded by electrons arranged in energy levels.

particle	relative charge	relative mass	location
proton	+1	1	nucleus
neutron	0	1	nucleus
electron	–1	very small	energy levels

- The number of protons in the nucleus of an atom is its atomic number.
- The number of electrons in an atom is the same as the number of protons.
- Atoms of the same element have the same atomic number.
- Atoms of different elements have different atomic numbers.
- Atoms are arranged in order of increasing atomic number in the periodic table.

Electron arrangement

- Electrons occupy energy levels. The lowest energy level is filled first.
- The electronic structure of an atom shows how its electrons are arranged in the different energy levels. For example, the electronic structure of sodium is 2,8,1. It shows that two electrons occupy the lowest energy level, then eight in the next energy level, and one electron in the highest energy level.
- The number of occupied energy levels is the same as the period number (row in the periodic table). The number of electrons in the highest energy level is the same as the group number (column in the periodic table).

Ions

- Ions are charged particles formed when atoms lose or gain electrons. The electronic structure of ions is the same as the structure of a noble gas.
- Metal atoms lose electrons to form positive ions. For example, the group 1 elements form ions with a single positive charge.
- Non-metal atoms gain electrons to form negative ions. For example, the group 7 elements form ions with a single negative charge.

Ionic compounds

- In chemical reactions, metal atoms can transfer electrons to non-metal atoms. Oppositely charged ions form when this happens.
- Ionic bonds are electrostatic forces of attraction between oppositely charged ions.
- Ionic compounds consist of very many ions arranged in a regular lattice. They form a giant structure held together by ionic bonds.

- A lot of energy is needed to overcome the very many strong ionic bonds in an ionic compound. So ionic compounds have high melting points and boiling points.

Simple molecules

- A covalent bond is a pair of electrons shared between two atoms.
- Covalent bonds form between non-metal atoms.
- Simple molecules consist of a small number of atoms joined together by covalent bonds. H_2, O_2, Cl_2, HCl, H_2O, NH_3 and CH_4 exist as simple molecules.
- There are weak forces attracting simple molecules towards each other. Little energy is needed to overcome these forces, so substances consisting of simple molecules have low melting points and boiling points.

Giant molecules

- Giant molecules consist of very many atoms joined together by covalent bonds to form a giant structure. Diamond, graphite and silica (silicon dioxide) exist as giant molecules.

diamond	graphite
Each carbon atom covalently bonded to four other carbon atoms.	Each carbon atom covalently bonded to three other carbon atoms.
Regular lattice of carbon atoms with strong covalent bonds makes diamond very hard.	Layers of carbon atoms with weak forces between them. Layers can slide over each other, so graphite is soft and slippery.

- A lot of energy is needed to overcome the very many strong covalent bonds in a giant molecule. So substances consisting of giant molecules have high melting points and boiling points.

Metals

- Metals can be bent and shaped without breaking because they have layers of atoms that can slide over each other.

Higher Tier

- Metals consist of a giant structure of metal ions in a regular arrangement, with free electrons. The metal ions are strongly attracted to the electrons.
- Graphite and metals are good conductors of heat and electricity because they have free electrons.

Isotopes

- Atoms are represented by chemical symbols like this: $^{23}_{11}Na$. The top number is the mass number (the number of protons and neutrons). The bottom number is the atomic number (number of protons).
- Isotopes are atoms of an element with different numbers of neutrons. They have the same atomic number but different mass numbers.

Relative atomic mass and relative formula mass

- The relative atomic mass of an element is the mean mass of its isotopes compared with the mass of a $^{12}_{6}C$ atom. It has the symbol A_r.
- The relative formula mass of a substance is all of the A_r values of the atoms in its formula added together. It has the symbol M_r.
- The M_r of a substance in grams is one mole of the substance.

Chemical calculations

- The percentage mass of an element in a compound can be calculated using its A_r value, and the M_r value and formula of the compound.
- The atom economy of a process is the percentage mass of useful product compared to the total mass of products or reactants.

Higher Tier

- The expected yield of a product can be calculated using the balanced symbol equation and the relative formula masses of the substances involved.
- The percentage yield is the actual yield compared to the expected yield, shown as a percentage.
- The concentration of a solution is given in moles per cubic decimetre.
- Equal volumes of gases at the same temperature and pressure contain the same number of molecules.

Rates of reaction

- The rate of a reaction can be found by measuring the rate that a reactant disappears, or the rate that a product appears.
- Reactions happen when particles collide with enough energy, called the activation energy.
- Reactions go faster if you increase the temperature, the concentration of a dissolved reactant, the pressure of a reacting gas or the surface area of a reacting solid. Catalysts also increase the rate of reactions.

Reversible reactions

- In a reversible reaction, the products can break down to form the reactants again.
- Exothermic reactions give out energy to the surroundings: the reaction mixture gets hot. Endothermic reactions take energy in from the surroundings: the reaction mixture gets cold.
- If the forward reaction in a reversible reaction is exothermic, the reverse reaction is endothermic.

Higher Tier

- If a reversible reaction happens in a closed container it reaches equilibrium. The forward and reverse reactions are still happening but at the same rate, so the concentrations of reactants and products stays the same.
- If the temperature is increased, the position of equilibrium moves in the direction of the endothermic reaction.
- If the pressure is increased in an equilibrium involving reacting gases, the position of equilibrium moves in the direction of the fewest molecules of gas.

The Haber process

- Ammonia is made from nitrogen and hydrogen in the Haber process. Nitrogen comes from the air. Hydrogen comes from reacting natural gas or coal with steam.
- The reaction is reversible: $N_2 + 3H_2 \rightleftharpoons 2NH_3$. The conditions used are 450 °C, 200 atmospheres pressure. An iron catalyst is used.

Higher Tier

- The temperature is a compromise: cold enough to get a reasonable yield of ammonia and hot enough to get a reasonable rate of reaction.

Electrolysis

- Ionic substances conduct electricity when they are molten or dissolved in water because their ions are free to move.
- Electrolysis is the breaking down of a compound to form elements using an electric current.
- Positively charged ions (hydrogen or metal ions) move to the negative electrode. They gain electrons and are reduced. If the metal involved is more reactive than hydrogen, hydrogen gas is formed at the electrode instead of the metal.
- Negatively charged ions (non-metal ions) move to the positive electrode. They lose electrons and are oxidised.

Acids, bases and salts

- Metal oxides and metal hydroxides are bases. Bases neutralise acids. Bases that dissolve in water are also called alkalis.
- Acids produce hydrogen ions H^+ and alkalis produce hydroxide ions OH^-.
- The equation for neutralisation is $H^+(aq) + OH^-(aq) \rightarrow H_2O(l)$.
- Different combinations of base and acid produce different salts. The first part of the salt's name comes from the metal in the base (or 'ammonium' if ammonia is used). The second part comes from the acid: sulfate from sulfuric acid, chloride from hydrochloric acid, and nitrate from nitric acid.
- Different salts are produced in different ways, depending on whether the salt is insoluble or not, and the reactivity of the metal in the salt.

Physics

Motion

- Speed tells us the rate at which an object moves. Speed is measured in m/s.
- You can find the speed of an object if you know how far it travels in a given amount of time: speed = distance ÷ time, where speed is measured in m/s, distance in m, time in s. The slope of a graph of distance against time represents the speed of an object.
- The velocity of an object is its speed in a certain direction. Velocity is measured in m/s.
- The area under a graph of velocity against time represents the distance travelled, and its slope represents acceleration.
- Acceleration is the rate of change of velocity. It is measured in m/s^2.

Forces

- The resultant force on an object is a force that would have the same effect as all the forces on an object acting together.
- If the resultant force on an object is zero, the object will stay still if it is not moving. It will carry on moving in the same direction and speed if it is already moving.
- If the resultant force on an object is not zero, the object will move in the direction of the resultant force if it is not moving. It will change direction and/or speed if it is already moving – it will accelerate.
- Force = mass × acceleration, where force is measured in N, mass in kg, acceleration in m/s^2.
- Weight = mass × gravitational field strength, where weight is measured in N, mass in kg, gravitational field strength in N/kg.
- The Earth's gravitational field strength is about 10 N/kg, so a 1 kg mass weighs 10 N.

Car stopping distances

- The stopping distance of a car depends upon the thinking distance and braking distance.
- The thinking distance is the distance travelled during the driver's reaction time. It increases if the driver is tired, or under the influence of drugs or alcohol.
- The braking distance is the distance travelled between applying the brakes and the car stopping. It increases if the brakes or tyres are worn, if the road is slippery, or if the car is heavily loaded.
- The thinking distance and braking distance increase as the speed of the car increases.

Work and energy

- Work and energy are both measured in joules, J.
- Work = force × distance, where work is measured in J, force in N, distance in m.
- The kinetic energy of a moving object depends on the object's mass and speed.

Higher Tier

- Kinetic energy = ½ × mass × (speed)², where kinetic energy is measured in J, mass in kg, speed in m/s.

Momentum

- A change in momentum happens when a force acts on an object that can move or is moving.
- momentum = mass × velocity where momentum is measured in kg m/s, mass in kg, velocity in m/s.
- The total momentum stays the same in an explosion or collision.

Higher Tier

- force = change in momentum ÷ time taken for change, where force is measured in N, change in momentum in kg m/s, time in s.

Static electricity

- Insulating materials can become charged when rubbed together because electrons move from one substance to the other.
- Objects with the same charge (both positive or both negative) repel.
- Objects with opposite charges (positive and negative) attract.
- Static electricity is used in photocopiers, laser printers and electrostatic precipitators in chimneys.
- Static electricity can be dangerous if there is a large potential difference; the charge is discharged to earth through a conductor.

Electrical circuits

- Electric current flows through a component when a potential difference is applied across it.
- Potential difference = current × resistance where potential difference is measured in V, current in A, resistance in Ω.
- Potential difference is measured using a voltmeter in parallel with the component, and current is measured with an ammeter in series with the component.
- Where components are in series, the total resistance is the individual resistances added together. The current is the same in each one, the total potential difference is shared between them.
- Where components are in parallel, the total current in the circuit is the individual currents added together, and the potential difference is the same in each one.

Mains electricity

- Batteries supply direct current d.c. in which the direction of the current stays the same.
- The mains supply is alternating current a.c. – it constantly changes direction.

- The mains supply has a frequency of 50 Hz and is about 230 V.
- The three-pin plug has several safety features, including an insulated case, a fuse, a cable grip and colour-coded wiring.
- Fuses and circuit breakers switch off a circuit if the current becomes too high because of an electrical fault.

Power
- Power = energy transformed ÷ time, where power is measured in W, energy in J, time in s.
- Power = current × potential difference, where power is measured in W, current in A, potential difference in V.

Higher Tier

Energy and charge
- The unit of electrical charge is the coulomb, C.
- Energy transformed = potential difference × charge, where energy is measured in J, potential difference in V, charge in C.
- Charge = current × time, where charge is measured in C, current in A, time in s.

Atomic structure
- The number of protons in the nucleus of an atom is its atomic number.
- The number of protons and neutrons in the nucleus of an atom is its mass number.
- Atoms of the same element have the same atomic number but atoms of different elements have different atomic numbers.
- Atoms that gain or lose electrons form charged particles called ions.
- Isotopes are atoms of an element with the same number of protons and electrons but different numbers of neutrons. Their atomic number is the same but their mass numbers are different.

Radioactive decay
- Some isotopes are unstable. Their nuclei decay and give off radiation.
- Alpha particles are identical to helium nuclei. They consist of two protons and two neutrons.
- When a nucleus gives off an alpha particle, its atomic number goes down by two and its mass number goes down by four. The nucleus of a different element is formed.
- Beta particles are high energy electrons given off from the nucleus. A neutron becomes a proton when a beta particle is given off.
- When a nucleus gives off a beta particle, its atomic number goes up by one and its mass number stays the same. The nucleus of a different element is formed.
- Radioactive isotopes decay all around us all the time. This is background radiation.
- Natural sources of background radiation include cosmic rays from space, radon gas from the ground, and radiation in food and drink.
- Artificial sources of background radiation include medical X-rays and radiation from nuclear waste.

Fission and fusion
- Nuclear fission is a change in which a nucleus splits to form smaller nuclei.
- Nuclear power stations rely on nuclear fission, usually with uranium-235 or plutonium-239 as their fuel.
- The nuclei of uranium-235 or plutonium-239 split into two smaller nuclei when a neutron hits them. They give off energy and two or three more neutrons when this happens. These can split even more nuclei, and a chain reaction happens.
- Nuclear fusion is a change in which nuclei join to form larger nuclei.
- Stars like our Sun rely on nuclear fusion to release energy from elements.